FLORA ZAMBESIACA

Flora terrarum Zambesii aquis conjunctarum

VOLUME SEVEN: PART TWO

FLORA ZAMBESIACA

MOZAMBIQUE
MALAWI, ZAMBIA, ZIMBABWE
BOTSWANA

VOLUME SEVEN: PART TWO

Edited by
E. LAUNERT

on behalf of the Editorial Board:

E. A. BELL
Royal Botanic Gardens, Kew

E. LAUNERT
British Museum (Natural History)

E. J. MENDES
*Centro de Botânica, Instituto de Investigação
Científica Tropical, Lisboa*

Published by the Managing Committee on behalf of
the contributors to Flora Zambesiaca
1985

Printed in Great Britain by Clark Constable,
Edinburgh, London, Melbourne

ISBN 0 565 00952 4

CONTENTS

LIST OF FAMILIES INCLUDED IN VOLUME VII, PART 2

111. APOCYNACEAE

By A. J. M. Leeuwenberg & F. K. Kupicha et al.*

Trees, shrubs, lianas, or sometimes herbs, mostly with white sticky latex. Leaves opposite, whorled, or less often alternate, simple, pinnately veined, entire. Stipules usually absent. Flowers bisexual, mostly actinomorphic, 5- or rarely 4-merous. Calyx often with colleters inside. Corolla tubular, sometimes with a corona; lobes contorted or occasionally (not in F.Z. area) valvate. Stamens inserted on the corolla; filaments free from each other or exceptionally (not in F.Z. area) united in a tube, often very short, frequently continued downwards as ridges at the corolla inside; anthers frequently triangular, connivent over and often coherent with the stigma, 2-celled, often partly sterile, sometimes with apical appendages; cells parallel, discrete, dehiscent throughout by a longitudinal slit. Pollen granular. Ovary superior or sometimes partly inferior, 1-celled and with 2 parietal placentas, 2-celled and with an axile placenta in each cell, or composed of 2, rarely more, separate or at the base partly united carpels each with an adaxial placenta; ovules 2 to many; style one, often split at the base when carpels more or less separate; stigma composed of a large variously shaped part, usually called the clavuncula, with laterally and/or basally the receptive zone, which is — if stigma coherent with the anthers — below the level of coherence, and a small apical usually sterile (also in F.Z.) so-called stigma. Disk annular, cupular, composed of separate glands, or absent. Fruit entire or consisting of two, rarely more, separate or partly united carpels, baccate, drupaceous, or follicular. Seeds in dry fruits often winged or with a coma, mostly with endosperm and a large embryo.

A family of c. 180 genera and about 1700 species, mainly in the tropics. Species of several genera are reputed for their medicinal properties (e.g. *Hunteria, Pleiocarpa, Voacanga, Tabernaemontana, Carvalhoa, Schizozygia, Catharanthus, Holarrhena, Rauvolfia, Strophanthus* and *Funtumia*).

1. Leaves alternate or fascicled; stems succulent - - - - - - - - 2
 – Leaves opposite, sometimes in some pairs subopposite, or whorled - - - 3
2. Stipules minute or absent; anthers produced into long hairy appendages   21. **Adenium**
 – Stipules transformed into rigid spines; anthers with only a short terminal appendage
                                                                20. **Pachypodium**
3. Plants armed with paired straight, usually branched spines; branches repeatedly dichotomously branched - - - - - - - - - - - -1. **Carissa**
 – Plants unarmed, dichotomously branched or not - - - - - - - - 4
4. Corolla lobes overlapping to the left - - - - - - - - - 5
 – Corolla lobes overlapping to the right - - - - - - - - - 22
5. Carpels completely fused, fruits fleshy, indehiscent - - - - - - 6
 – Carpels free, connate at the base, or sometimes (*Callichilia orientalis*) up to $\frac{9}{10}$ fused; fruit fleshy or dry - - - - - - - - - - - - - - - 12
6. Tree or shrub without tendrils; corolla lobes much shorter than the tube, less than 1·5 times as long as wide; fruit 1–2-seeded - - - - - - - 2. **Acokanthera**
 – Climber with curled tendrils or sometimes (*Chamaeclitandra*) sparingly branched shrub; corolla lobes shorter or longer than the tube, usually at least twice as long as wide; fruit usually with 3 or more seeds - - - - - - - - - - - 7
7. Inflorescence of elongate, branched, terminal panicles - - - - - - 8
 – Inflorescence contracted, short and often clustered, terminal or axillary - - - 9

---

* This contribution was undertaken by a team of seven authors, some based on monographic revisions of their own. The authorship is shared as follows: A. J. M. Leeuwenberg (family diagnosis, key to the genera, *Carvalhoa, Tabernaemontana, Voacanga,* and *Wrightia,* and notes on cultivated species; F. K. Kupicha (*Acokanthera, Alafia, Ancylobotrys, Aphanostylis, Carissa, Chamaeclitandra, Dictyophleba, Hunteria, Landolphia, Mascarenhasia, Oncinotis, Pachypodium, Pleiocarpa,* and *Saba*); M. M. Barink (*Pleioceras* and *Schizozygia*); H. J. Beentje (*Callichilia* and *Strophanthus*); A. P. M. de Kruif (*Baissea, Holarrhena,* and *Rauvolfia*); A. C. Plaizier (*Adenium, Catharanthus* and *Diplorhynchus*); H. J. C. Zwetsloot (*Funtumia*). Dr. F. K. Kupicha undertook this work in her capacity as Krukoff Curator of African Botany. The final text was revised and compiled by the Editor of this volume.

8. Anthers not keeled; ovary and fruit glabrous; stipules interpetiolar, soon caducous
6. **Dictyophleba**
  − Anthers keeled along the back; ovary and fruit pubescent; stipules absent
7. **Ancylobotrys**
9. Length of the anthers almost $\frac{1}{2}$ or more of the length of the corolla tube; clavuncula ring-shaped - - - - - - - - - - - - 5. **Aphanostylis**
  − Length of the anthers up to one third the length of the corolla tube, frequently less -   10
10. Wall of the corolla tube thickened above the anthers; anthers c. 0·75 mm. long; inflorescences axillary or axillary and terminal - - - - - - - 11
  − Wall of the corolla tube not thickened above the anthers; anthers c. 2·75 mm. long and $\frac{1}{10}-\frac{1}{6}$ of the tube; only terminal inflorescence developed - - - - - 8. **Saba**
11. Rhizomatous shrub with straight stems; ovules in 2 series of (1) 2 (3); stigma 0·02–0·07 mm. long - - - - - - - - - - - - 4. **Chamaeclitandra**
  − Climber with curled tendrils; ovules in 4–18 series of 4–13; stigma 0·15–1·85 mm. long
3. **Landolphia**
12. Stems subherbaceous, not exceeding 1·5 m.; with showy pink or white trumpet-shaped flowers; follicles small, green, with dark brown subglobose seeds without coma
17. **Catharanthus**
  − Stems woody; shrubs, trees or climbers - - - - - - - - - 13
13. Anthers not sagittate, fertile to the base; variously, but not strictly dichotomously branched shrubs or trees - - - - - - - - - - - 14
  − Anthers sagittate or auriculate, often sterile towards the base (see *Tabernaemontana*); usually strictly dichotomously branched shrubs or trees with inflorescences in the forks
17
14. Fruit composed of two dry follicles; seeds winged; leaves long-petiolate; petiole about $\frac{1}{3}$ of the lamina; inflorescence terminal; corolla-tube 2–3 mm. long   - 16. **Diplorhynchus**
  − Fruit fleshy, indehiscent; petiole less than $\frac{1}{4}$ of the lamina; inflorescence terminal or not; corolla tube usually more than 3 mm. long - - - - - - - 15
15. Leaves in most branchlets whorled; calyx without colleters - - - - - 16
  − Leaves opposite; calyx with colleters; inflorescence terminal or occasionally axillary, paniculate, lax and with opposite branches; fruit a berry   - - - 9. **Hunteria**
16. Inflorescence usually distinctly pedunculate (not in *R. nana*) and with whorled branches; disk present; fruit a drupe - - - - - - - - - 19. **Rauvolfia**
  − Inflorescence fasciculate, axillary; disk absent; fruit a berry - - - 10. **Pleiocarpa**
17. Corolla with a lobed or partite corona; fruit of dry slender follicles; seed with a terminal coma, directed towards the base of the fruit - - - - - - - - 18
  − Corolla without corona; fruit fleshy, of subglobose to follicular carpels, dehiscent or not; seeds without coma - - - - - - - - - - - 19
18. Lamina of leaves 2·6–3·5 times as long as wide, distinctly hairy; calyx 2–3 mm. long; corolla dark violet with yellow; tube 6–9 mm. long - - - - - - 24. **Pleioceras**
  − Lamina of leaves 3–6·5 times as long as wide, glabrous or only hairy beneath at the base of the midrib, pubescent and sometimes with minute scattered hairs above; calyx 5–6 mm. long; corolla creamy; tube 4–5 mm. long - - - - - - - 23. **Wrightia**
19. Sepals united for at least $\frac{2}{3}$ of their length, deciduous; corolla tube slightly shorter to slightly longer than the calyx; style and stigma shed with the corolla; carpels subglobose, dark green and very pale green-spotted, dehiscent - - - - - - 11. **Voacanga**
  − Sepals free or nearly so, persistent on fruit; corolla tube at least twice as long as the calyx; style and stigma persistent when the corolla is shed; carpels smooth, sometimes dotted or verrucose, green or orange - - - - - - - - - - - 20
20. Carpels united for $\frac{2}{3}-\frac{9}{10}$ of their length, ellipsoid or nearly so; sepals 2–3 times as long as wide and rounded at the apex; corolla tube slenderly cylindrical   - 15. **Callichilia**
  − Carpels free or nearly so; sepals about as long as wide, or if up to 1·7 times as long as wide then acute or acuminate and follicles long and slender - - - - - 21
21. Leaves membranous when dry; corolla with many red longitudinal lines at the base of the lobes; tube campanulate; follicles slender, orange, 3–6 × 0·8–1 cm.-   - 13. **Carvalhoa**
  − Leaves coriaceous, also when fresh; corolla without red lines; tube almost cylindrical to more or less bottle-shaped; follicles subglobose or ellipsoid, much wider
12. **Tabernaemontana**
22. Repeatedly dichotomously branched shrub or tree with 2 short few-flowered inflorescences in the forks, flowers small; carpels elliptic, about as long as wide; seeds without coma
14. **Schizozygia**
  − Variously, but not dichotomously, branched woody plants; inflorescences terminal or axillary; carpels dry follicles, many times as long as wide; seeds with coma   - 23
23. Anthers not caudate, fertile to the base; inflorescence many-flowered, $\pm$ congested; corolla white, with a slender tube and a widely spreading limb; follicles slender; seeds not rostrate
18. **Holarrhena**
  − Anthers sagittate or auriculate, sterile towards the base; corolla white or not; follicles slender or not; seeds rostrate or not - - - - - - - - - 24
24. Corolla white with red (see also *Baissea wulfhorstii*), generally turning yellow with purple, with paired corona appendages alternating with the (frequently long-caudate) corolla

lobes; tube usually more than 15 mm. long; seeds produced into a long plumose apical rostrum, directed towards the apex of the fruit -   -   -   -   22. **Strophanthus**
- Corolla usually evenly coloured (not in *Baissea wulfhorstii*) without or with (*Oncinotis*) solitary corona appendages; tube up to 12 mm. long; seed — if produced into an apical rostrum — directed towards the base of the fruit (*Funtumia*) -   -   -   -   - 25
25. Inflorescence paniculate, axillary; corolla with solitary corona-appendages; tube about 3 mm. long; seeds not rostrate; large liana -   -   -   -   - 28. **Oncinotis**
- Inflorescence corymbose or nearly so, terminal and often at the same time axillary; corolla-tube at least 4 mm. long; trees, shrubs, or climbers -   -   -   -   -   - 26
26. Corolla tube narrowed at the mouth; disk present or absent; seeds rostrate or not - 27
- Corolla tube widened at the mouth; disk present; seeds not rostrate; climbers
29. **Baissea**
27. Inflorescence terminal only; climber or climbing shrub; disk absent; follicles often slightly torulose, very long and slender -   -   -   -   -   -   -   - 27. **Alafia**
- Inflorescences terminal and axillary; erect shrub or tree; disk present; follicles stiff, woody or grooved, not torulose -   -   -   -   -   -   -   -   -   - 28
28. Sepals glabrous or puberulous; corolla lobes obliquely ovate to narrowly oblong, glabrous to puberulous, obtuse or acute; seeds with an apical rostrum directed towards the base of the fruit -   -   -   -   -   -   -   -   -   -   - 25. **Funtumia**
- Sepals thinly appressed-pilose; corolla lobes almost rhombic, densely hairy, acute or acuminate; seeds not rostrate -   -   -   -   -   - 26. **Mascarenhasia**

Key to cultivated species

Many exotic species of the *Apocynaceae* are grown for ornament in the F.Z. area, *Allamanda cathartica, A. schottii* (= *A. neriifolia* Hook.), *Plumeria rubra* ("Frangipani") and *Thevetia peruviana* being the most frequent; with the exception of *Nerium oleander* (= *N. indicum* Mill.), a species indigenous to N. Africa and S. Europe they are all natives of other continents. *Allamanda cathartica* L., *A. schottii* Pohl, *A. violacea* Gardn. & Field, *Mandevilla laxa* (Ruiz & Pav.) Woods., *Plumeria alba* L., *P. rubra* L., *Thevetia peruviana* (Pers.) K. Schum. and *T. thevetioides* (Kunth) K. Schum. are American whereas *Alstonia macrophylla* Wall. ex G. Don, *A. venenata* R. Br., *Alyxia ruscifolia* A. Cunn., *Beaumontia grandiflora* (Roxb.) Wall., *Tabernaemontana divaricata* (L.) R. Br. ex Roem & Schult. (= *Ervatamia coronaria* (Jacq.) Stapf), *Trachelospermum jasminoides* (Lindl.) Lem., and *Vallaris solanacea* (Roth) O. Kuntze (= *V. heynei* Spreng.) are Asian; *Vinca major* L. is native to Europe.

1. Plants herbaceous with opposite almost cordate leaves on creeping vegetative and short erect flowering stems; corolla blue-violet -   -   -   -   -   - *Vinca major*
- Plants woody -   -   -   -   -   -   -   -   -   -   -   -   - 2
2. Leaves alternate, narrowly or very narrowly elliptic -   -   -   -   -   - 3
- Leaves opposite or whorled -   -   -   -   -   -   -   -   -   - 6
3. Leaves very narrowly elliptic, up to 1cm. wide; flowers large, yellow; branchlets thin; evergreen tree or shrub -   -   -   -   -   -   -   -   -   - 4
- Leaves at least 3 cm. wide; flowers pinkish or nearly so, fleshy; branchlets thick, more or less succulent; shrub or small tree, often leafless, but still flowering, in the dry season   5
4. Corolla lobes erect, remaining contorted and overlapping at anthesis; leaves always glabrous beneath -   -   -   -   -   -   -   - *Thevetia peruviana*
- Corolla lobes spreading, separating at anthesis; leaves often hairy beneath
Thevetia thevetioides
5. Leaves glabrous beneath, flat, with flat margin -   -   -   - *Plumeria rubra*
- Leaves hairy beneath, often more or less bullate, with the margin revolute *Plumeria alba*
6. Leaves whorled; usually trees or shrubs; flowers variously coloured -   -   - 7
- Leaves opposite; usually climbers; flowers usually white -   -   -   - 13
7. Leaves up to 25 mm. long, sharply pointed at apex; flowers about 5–7 mm. long
Alyxia ruscifolia
- Leaves larger, usually more than 4 cm. long, not sharply pointed; flowers usually at least 2 cm. long -   -   -   -   -   -   -   -   -   -   -   -   - 8
8. Leaves strictly ternate, stiffly coriaceous, with a minute reticulate venation beneath; flowers usually pink, often double; anthers with long hairy appendages -   - *Nerium oleander*
- Leaves ternate or quaternate, herbaceous when fresh, venation not reticulate; flowers variously coloured; anthers without appendages -   -   -   -   -   - 9
9. Trees or shrubs; flowers small; tube nearly cylindrical, up to 30 × 2 mm.; fruit of slender long follicles, smooth; seeds with a coma at both ends -   -   -   -   - 10
- Climbers or shrubs; flowers much larger; tube ± infundibuliform, at least 50 mm. long; fruit subglobose, prickly -   -   -   -   -   -   -   -   -   - 11
10. Branchlets subquadrangular; leaves pubescent all over or only along the main veins beneath; secondary veins more than 8 mm. from each other; corolla tube 4·5–6 mm. long; lobes as long as the tube or slightly longer -   -   -   - *Alstonia macrophylla*
- Branchlets terete; leaves glabrous or pubescent on both sides; secondary veins more than 10 per cm.; corolla 20–30 mm. long; lobes much shorter than the tube *Alstonia venenata*

398   III. APOCYNACEAE

11. Corolla mauve - - - - - - - - - - - *Allamanda violacea*
 – Corolla yellow, often with some red - - - - - - - - - - 12
12. Corolla tube widely infundibuliform; the wide part about as long as the narrow; corolla usually not red-striped; climber with white latex - - - - *Allamanda cathartica*
 – Corolla tube narrowly infundibuliform to almost cylindrical; the wide part about 3 times as long as the narrow, inside in throat with red longitudinal lines, clearly longitudinally veined outside when dry; shrub usually(?) with clear sap - - *Allamanda schottii*
13. Repeatedly dichotomously branched shrub or small tree with inflorescences in the forks; flowers usually double and sweet-scented ("coffee rose") *Tabernaemontana divaricata*
 – Climbers, variously, but not dichotomously branched - - - - - - - 14
14. Flowers very large; corolla tube about 10 cm. long; sepals more than 30 mm. long; follicles about 20 × 2·5 cm. - - - - - - - - *Beaumontia grandiflora*
 – Flowers much smaller; corolla tube less than 6 cm. long; sepals up to 12·5 mm. long   15
15. Corolla with a very wide limb; tube 5–10 mm. long; follicles 8–15 cm. long; leaves cuneate or rounded at the base - - - - - - - - - - - 16
 – Corolla more or less infundibuliform; tube 25–55 mm. long; follicles 25–40 × 0·5 cm.; leaves more or less cordate at the base - - - - - - - *Mandevilla laxa*
16. Corolla tube slightly widened around the included anthers; follicles slender, about 5 mm. in diam. - - - - - - - - - *Trachelospermum jasminoides*
 – Corolla tube not widened; anthers exserted; follicles wider, 1·5–3·5 cm. in diam.
    Vallaris solanacea

H. N. Biegel, A Checklist of ornamental plants used in Rhodesian parks and gardens (Rhod. Agric. Journal, Res. Rep. No. 3, 1977) lists the following species not included in the key above: *Acokanthera friesiorum* Markgraf, *A. oppositifolia* (Lam.) Codd (see p. 405), *A. schimperi* (A. DC.) Schweinf. (see p. 406), *Adenium multiflorum* Klotzsch (see p. 465), *Carissa edulis* (Forssk.) Vahl (see p. 399), *Catharanthus roseus* (L.) G. Don (see p. 454), *Holarrhena pubescens* (Buch.-Ham.) Wall. ex G. Don (see p. 456), *Mascarenhasia arborescens* A.DC. (see p. 487), *Pachypodium saundersii* N.E.Br. and *Rauvolfia caffra* Sond. (see p. 460).

1. CARISSA L.

Carissa L., Syst. Nat., ed. 12, **2**: 189 (1767); Mant. Pl.: 7, 52 (1767), *nom. conserv.* — Markgraf in Notizbl. Bot. Gart. Berlin **15**: 455 (1942). — Pichon in Mém. Mus. Nation. Hist. Nat., N.S. **24**: 130 (1948) *pro parte excl.* Sect. *Acokanthera* (G. Don) Pichon.
Carandas Adans., Fam. Pl. **2**: 171, 532 (1763), *nom. rej.*
Arduina Miller ex L., Syst. Nat., ed. 12, **2**: 136, 180 (1767); Mant. Pl.: 7, 52 (1767).
Antura Forssk., Fl. Aegypt.-Arab.: 63 (1775).
Jasminonerium Kuntze, Rev. Gen. Pl. **2**: 414 (1891).

Shrubs or small trees, much branched, sometimes scandent, with simple or forked spines. Stipules absent. Flowers in terminal corymbs or 1–5-flowered cymes. Calyx lobes imbricate, free to base. Corolla hypocrateriform; tube cylindrical, ± straight, hairy within; lobes contorted, overlapping either to the left or the right, of various shapes, much shorter than to longer than the tube. Stamens inserted at the middle or towards the top of the corolla tube; anthers subsessile, without carina, glabrous. Ovary glabrous, bicarpellate, with 1–4 or rarely many ovules per loculus; style slender; stigma below or reaching to androecium, bilobed, glabrous; clavuncle often inconspicuous. Fruit an ellipsoid or ovoid berry containing (1)2–8(∞) seeds. Seeds discoid or compressed-ellipsoid, velutinous.

A genus of c. 20 species distributed in Asia, Australia, Madagascar, the Mascarenes and Africa, with 7 species in Africa. The two African species not represented in the F.Z. area are both endemic to S. Africa.

The spines of *Carissa* represent modified inflorescences (Markgraf, loc. cit.). Each pair terminates its branch, and continued upward growth is by shoots arising in the axils of one or both of the leaves at right angles to the spines. This growth-pattern results in the much-branched, zig-zag appearance typical of *Carissa*.

1. Spines almost always simple, rarely furcate; young twigs densely pubescent or puberulous; corolla pink to crimson on the outside, with lobes overlapping to the right   - 1. *edulis*
 – Spines furcate or bifurcate, rarely simple or absent; young twigs glabrous or puberulous; corolla white outside, sometimes tinged with pink or green, with lobes overlapping to the left   2
2. Flowers tetramerous; corolla lobes broadly elliptic, much shorter than corolla tube; leaf lamina usually elliptic, the margin faintly crenate, the surfaces having a silky appearance   2. *tetramera*

- Flowers pentamerous; corolla not as above; leaf lamina usually ovate, the margin smooth, the surfaces not silky - - - - - - - - - - - - - 3
3. Spines usually bifurcate; corolla tube 11–18·5 mm. long; corolla lobes elliptic, longer than tube or rarely slightly shorter - - - - - - - - - - 4
- Spines usually furcate; corolla tube 4·5–11 mm. long; corolla lobes ovate-acuminate, ⅓ as long as to equalling tube - - - - - - - - - - - 5
4. Young stems glabrous, greenish, strongly rugose; leaves 2·4–7 × 1·7–5·2 cm.; calyx with tiny dark scales at base on inner side; fruits 3–5 cm. long, many-seeded - - 3. *macrocarpa*
- Young stems crispate-puberulous, reddish-brown, ± smooth; leaves 1·5–3·5(5) × 0·7–2(3) cm.; calyx without scales; fruits c. 1·5 cm. long, 8-seeded - - - - 4. *praetermissa*
5. Leaf-lamina thickly coriaceous, ovate or subcircular, almost always less than twice as long as broad; spines numerous, stout, with axis 20–38 mm. and branches (3)15–46 mm. long; stamens inserted near the top of the corolla tube; stigma apex attaining or surpassing level of anther bases - - - - - - - - - - 5. *bispinosa* subsp. *bispinosa*
- Leaf lamina thinly coriaceous, lanceolate, ovate or elliptic, almost always more than twice as long as broad; spines absent or relatively few, slender, with axis 2–13 mm. and branches 3–12 mm. long; stamens inserted at or near the middle of the corolla tube; stigma apex below level of anther-bases - - - - - - - - - 5. *bispinosa* subsp. *zambesiensis*

1. **Carissa edulis** (Forssk.) Vahl, Symb. Bot. **1**: 22 (1790). — Stapf in F.T.A. **4**, 1: 89 (1902). — Pichon in Mém. Inst. Sci. Madag., Sér. B, **2**: 127 (1949), *excl. syn. C. abyssinica* R. Br. *et C. africana* A. DC. — Dale & Greenway, Kenya Trees & Shrubs: 45 (1961). — Irvine, Woody Pl. Ghana: 616, fig. 120 (1961).— F. White, F.F.N.R.: 347 (1962). — H. Huber in F.W.T.A., ed. 2, **2**: 54 (1963). — Codd in Fl. Southern Afr. **26**: 251 (1963). — Fanshawe, Check List Woody Pl. Zambia: 9 (1973). — R. B. Drumm. in Kirkia **10**: 268 (1975). Type from Yemen.
  Antura edulis Forssk., Fl. Aegypt.-Arab.: cvi, 63 (1775). Type as above.
  Antura hadiensis G. F. Gmelin, Syst. Nat., ed. 13: 405 (1791), *nom. superfl. illegit.* Type as above.
  Arduina edulis (Forssk.) Spreng., Syst. Veg. **1**: 669 (1825). Type as above.
  Carissa dulcis Schumach. & Thonn. [ex Schumach., Beskr. Guin. Pl.: 146 (1827?)] in Kongel. Dansk. Vid Selsk. Naturvid. Math. Afh. **3**: 166 (1828). Type from Guinea.
  Carissa pubescens A. DC. in DC., Prodr. **8**: 334 (1844). Type from Senegal.
  Carissa tomentosa A. Rich., Tent. Fl. Abyss. **2**: 30 (1851). Type from Ethiopia.
  Carissa pilosa Schinz in Verh. Bot. Ver. Prov. Brand. **30**: 258 (1888). Type from Namibia.
  Jasminonerium edule (Forssk.) Kuntze, Rev. Gen. Pl. **2**: 415 (1891). Type as for *Carissa edulis.*
  Jasminonerium dulce (Schumach.) Kuntze, loc. cit. Type as for *Carissa dulcis.*
  Jasminonerium pubescens (A. DC.) Kuntze, loc. cit. Type as for *Carissa pubescens.*
  Jasminonerium tomentosum (A. Rich.) Kuntze, loc. cit. Type as for *Carissa tomentosa.*
  Carissa edulis var. *tomentosa* (A. Rich.) Stapf in F.T.A. **4**, 1: 90 (1902). Type as above.
  Carissa edulis var. *major* Stapf, loc. cit. Syntypes from Transvaal, Angola and Malawi: Manganja Hills, ix–xi.1861, *Meller* s.n. (K, lectotype); foot of Mt. Chiradzulu, 3.x.1859, *Kirk* s.n. (K, lecto paratype).
  Azima pubescens Suesseng. in Mitt. Bot. Staatss. München **1**: 334 (1953). Syntypes from Namibia.

Much-branched spreading or sarmentose shrub up to 5·5 m. high or more. Young twigs densely pubescent or puberulent, rarely glabrous, the surface smooth or very finely rugose; older branches with grey-brown flaking bark. Spines simple, very rarely furcate, 0·4–5·5(7) cm. long. Leaves moderately coriaceous, drying discolorous, brown or green, darker above; petiole 2·5–5 mm. long, pubescent or glabrous. Lamina (1·7)2·6–6·8 × (1)1·2–4·6 cm., ovate, elliptic, obovate-elliptic or subcircular, with apex acute or obtuse, with or without mucro, and base cuneate to rounded; upper surface glabrous to pubescent, midrib slightly impressed, other veins obscure; lower surface glabrous to pubescent (hairs most dense on midrib, here sometimes almost lanate), midrib and other veins slightly raised, venation easily visible or obscure. Cymes short-stalked, few- to many-flowered; flowers 5 (occasionally 6)-merous, sweetly scented, the corolla white inside and pink to crimson outside. Calyx 2–4 mm. long; lobes lanceolate-subulate, glabrous, pubescent or lanate. Corolla tube 9·5–20 mm. long, slightly wider in upper half and contracted at the mouth, glabrous or sparsely pilose without, pubescent within in upper half; corolla lobes ⅓–½ as long as tube, lanceolate, glabrous or sparsely puberulent near the mouth on upper surface, overlapping to the right. Stamens inserted near apex of tube so that the anthers reach to within 1–1·5 mm. of the mouth; anthers subsessile, 1·5–2 mm. long. Ovary c. 1 mm. long, rounded-conical, glabrous;

style glabrous, stigma reaching to base of anthers. Fruit 6–11(25) mm. long, plum-shaped, green turning red to blue-black, edible, 2–4-seeded.

Caprivi Strip. Lisikili, 24 km. E. of Katima Mulilo, 975 m., fl. 17.vii.1952, *Codd* 7106 (PRE). **Botswana.** N: Namatanga, near Linyanti R., st. 29.x.1972, *Biegel, Pope & Russell* 4100 (LISC; PRE; SRGH). **Zambia.** B: Senanga Distr., Kaunga near Mashi R., fl. 20.viii.1962, *Mubita & Reynolds* B164 (SRGH). N: Mbala (Abercorn) Distr., Lumi Marsh, Kawimbe, 1740 m., fl. 25.x.1962, *Richards* 16850 (BR; K; SRGH). W: Ndola Distr., fl. 26.ix.1947, *Brenan* 7962 (FHO; K). C: Lusaka Distr., Chalimbana area 3 km. S.E. of Annisdale, 1250 m., fl. & fr. 15.x.1972, *Strid* 2322 (FHO). E: Chadiza, 850 m., fl. 25.xi.1958, *Robson* 687 (BM; BR; K; LISC; PRE; SRGH). S: Mazabuka Distr., Chimundini near Kalomo, fl. 7.x.1955, *Gilges* 449 (LISC; PRE; SRGH). **Zimbabwe.** N: Lomagundi Distr., Matoroshanga, 1220 m., fl. 25.x.1959, *Leach* 9504 (LISC; SRGH). W: Bulawayo, 1370 m., fl. ix.1902, *Eyles* 23 (BM; SRGH). C: Marondera (Marandellas) Distr., 16 km. S. of Marondera, 1370 m., fr. 16.xii.1962, *Moll* 314 (SRGH). E: Mutare (Umtali) Distr., Odzani R., Umtasa Reserve, fl. 3.xii.1950, *Chase* 3234 (BM; BR; COI; LISC; SRGH). S: Chibi Distr., Sikanajena hills near Lundi R., st.xii.1955, *Davies* 1787 (K; PRE; SRGH). **Malawi.** N: Chitipa Distr., 32 km. S.E. of Chisenga, fl. 3.i.1977, *Pawek* (MAL). C: Dedza Distr., Chongoni Forest, fr. 22.v.1962, *Banda* 442 (FHO; K; LISC; MAL; SRGH). S: Blantyre, fr. immat. 1896, *Buchanan* 150 (K; PRE). **Mozambique.** Z: Alto Molócuè, 21 km. to Alto Ligonha, 600 m., fr. 29.xi.1967, *Torre & Correia* 16279 (C; LISC; LMA; MO; WAG). T: Tete, Marávia, Fingoè, near R. Luatize, fr. 28.vi.1949, *Andrada* 1681 (COI; LISC). MS: Chimoio, Gondola, Pindanganga, fl. 15.x.1945, *Simão* 587 (LISC).

Very widely distributed in Africa from Senegal to Somalia and from Sudan to the Transvaal and Namibia and also in Madagascar. Also occurring in Asia from Yemen to India and Thailand, and on islands of the Indian Ocean. In open savanna woodland, often on termitaria, and in riverine fringe vegetation.

The leaves of *C. edulis* vary from completely glabrous to densely pubescent, with every intermediate degree of hairiness. In the F.Z. area both glabrous and hairy specimens are known for nearly all parts of the species' range. In view of the continuous nature of the variation and absence of any pattern in its geographical distribution this character, although often striking, is not thought to be of taxonomic significance.

Pichon, loc. cit. dealt with this species over its whole range and gives a fuller synonymy.

2. **Carissa tetramera** (Sacleux) Stapf in F.T.A. **4**, 1: 91 (1902). —Dale & Greenway, Kenya Trees & Shrubs: 45 (1961).—Codd in Fl. Southern Afr. **26**: 252, fig. 37, 4 (1963).—R. B. Drumm. in Kirkia **10**: 268 (1975). — Compton, Fl. Swaziland: 439 (1976). Type from Zanzibar.

Arduina tetramera Sacleux in Journ. Bot., Paris **7**: 312 (1893). Type as above.

Low shrub 0·4 m. (or less) to 2 m. high. Young stems crispate-puberulous, very soon developing a pale grey flaking bark. Spines usually furcate with axis 2–24 mm. long and branches 3–23 mm., sometimes bifurcate; abortive flower sometimes present in the fork of spine branches. Leaves moderately coriaceous; petiole 1–2 mm. long. Lamina 1·6–6·3 × 0·4–3 cm., elliptic, lanceolate, ovate-elliptic or subcircular with apex acute to obtuse, mucronate and base rounded to cuneate; margin faintly crenate, flat or somewhat revolute; upper surface silky in appearance, glabrous or pubescent, midrib impressed, lateral nerves raised, indistinct to fairly conspicuous, apparently running right to the margin; lower surface silky, glabrous or pubescent, midrib and lateral nerves raised, the latter indistinct to fairly conspicuous. Flowers tetramerous, in sessile or shortly pedunculate few-flowered cymes, white, scented. Calyx 1–1·5 mm. long, of two larger and two smaller sepals; sepals ovate-triangular, ± free, dorsally glabrous or minutely puberulous, minutely ciliate. Corolla tube 6·5–11·5 mm. long, externally glabrous or puberulent, internally pubescent in upper ⅔ (density of indumentum variable). Corolla lobes 2–4 mm. long, broadly elliptic, lower surface glabrous or pubescent, upper surface subglabrous to densely villous near mouth of tube, indumentum (when present) extending into the throat; lobes overlapping to the left. Stamens inserted at or just above the middle of the corolla tube; anthers subsessile, 0·75–1·5 mm. long. Gynoecium glabrous; ovary c. 0·5 mm. long, subglobose; style and stigma 2–3 mm. long, stigma not reaching to anthers; clavuncle obscure. Fruit c. 1 cm. long, ellipsoid, red to black, edible, 4–8-seeded.

Zimbabwe. S: Beitbridge Distr., Chiturupazi, fl. 25.ii.1961, *Wild* 5387 (COI; K; PRE; SRGH). **Mozambique.** N: c. 11 km. from Angoche (António Enes) to Namaponda, c. 30 m., fl. & fr. 31.iii.1964, *Torre & Paiva* 11528 (C; EA; LISC; LMA; MO; SRGH; WAG). MS: Sofala Prov., Beira, 15 m., fl. 25.xii.1906, *Swynnerton* 1068 (BM; K). GI: Gaza, Caniçado, c. 17

km. from the point of Lagoa Nova to Aldeia da Barragem, fr. 17.vii.1969, *Correia & Marques* 918 (LMU). M: Goba, near R. Maiuana, fr. 4.xi.1960, *Balsinhas* 196 (COI; K; LISC; PRE; SRGH).

Also occurring in Kenya, Tanzania, Swaziland and S. Africa (eastern Transvaal and northern Natal). In open woodland.

3. **Carissa macrocarpa** (Ecklon) A. DC. in DC., Prodr. **8**: 336 (1844). — Codd in Fl. Southern Afr. **26**: 254 (1963). Type from S. Africa (Natal).

    Arduina macrocarpa Ecklon in S. Afr. Quart. Journ. **1**: 372 (1830). Type as above.

    Arduina grandiflora E. Mey., Comm. Pl. Afr. Austr.: 191 (1837). Type from S. Africa (Natal).

    Carissa grandiflora (E. Mey.) A. DC., tom. cit.: 335 (1844). — Hook. f. in Bot. Mag.: t. 6307 (1877). — Stapf in F.C. **4**, 1: 497 (1907). Type as above.

    Carissa africana A. DC., tom. cit.: 332 (1844). Type: Mozambique, Mozambique I., *Loureiro* s.n. (P).

    Jasminonerium grandiflorum (E. Mey.) Kuntze, Rev. Gen. Pl. **2**: 415 (1891). Type as for *Carissa grandiflora*.

    Jasminonerium africanum (A. DC.) Kuntze, loc. cit. Type as for *Carissa africana*.

Many-stemmed shrub 1–3·7 m. high. Young stems glabrous, greenish, strongly wrinkled into longitudinal ridges and channels; older twigs similar, with greenish or brownish, often glossy, surface. Spines usually bifurcate, the axis (2)5–22 mm., first branches (2)3–23 mm. and second branches (2)4–32 mm. long. Leaves thickly coriaceous, glabrous, often drying discolorous with dark grey to brown upper surface and pale green to brown lower surface; petiole 2–4 mm., glabrous. Lamina 2·4–7 × 1·7–5·2 cm., broadly to narrowly ovate or subcircular, the apex obtuse, mucronate, the base rounded; upper surface with midrib impressed and lateral nerves faintly raised, usually indistinct; lower surface with midrib prominent and lateral nerves level or ± impressed, very indistinct; margin revolute, not crenate. Flowers pentamerous, in sessile or shortly pedunculate terminal 1–few-flowered cymes, white, jasmine-scented, heterostylous. Calyx 2–4·5(7) mm. long; segments narrowly triangular to broader and somewhat foliaceous, glabrous, not ciliate; fleshy triangular dark-coloured scales c. 0·5 mm. long present, variable in number, shortly attached to base of sepals on their margins and inner surface. Corolla tube 11–18·5 mm. long, glabrous or pubescent on outer surface, sparsely to densely pilose within on the upper $\frac{2}{3}$. Corolla lobes 9–24 mm. long, broadly elliptic, glabrous or pubescent on upper and lower surfaces, overlapping to the left. Stamens inserted at the middle of the corolla tube; anthers subsessile, 1·25–2 mm. long (1·25 mm. and apparently non-functioning in long-styled flowers, 1·75–2 mm. in short-styled flowers). Gynoecium glabrous; ovary fusiform and 2–3·5 mm. long in long-styled flowers or conical and up to 1 mm. long in short-styled flowers; style and stigma 6–7 mm. long, the stigma reaching beyond the anthers, or only 2·5–3 mm.; stigma pubescent at apex, not obviously bifid. Fruit 3–5 cm. long when fresh, ovoid, pointed, red and edible when ripe, many-seeded. Seeds c. 6 × 4 mm.

Mozambique. N: Angoche (António Enes), fl. 20.x.1965, *Gomes e Sousa* 4877 (COI; K; PRE). Z: Pebane beach, near lighthouse, c. 10–20 m., fl. 12.i.1968, *Torre & Correia* 17101 (C; LISC; LMA). M: Inhaca I., fl. 23.vii.1956, *Mogg* 31283 (PRE; SRGH).

Also occurs in S. Africa (Natal and Cape Prov.). In coastal scrub, on sand dunes.

This species is grown in gardens in the F.Z. area, and escaped specimens are found outside its natural range. With its dimorphic flowers, *C. macrocarpa* is functionally dioecious; plants with short-styled flowers do, however, set occasional fruits (Codd, loc. cit.).

4. **Carissa praetermissa** Kupicha in Kew Bull. **36**: 47, fig. 1 (1981). TAB. 90. Type from Mozambique, GI: Inhambane, Bazaruto I., lighthouse hill, c. 50 m., fl. 20. x. 1958, *Mogg* 28448 (K, holotype; BM; LISC; LMU; SRGH).

Many-stemmed rhizomatous shrub 0·5–3 m. high. Young twigs thinly to densely crispate-puberulous, reddish-brown, the surface ± smooth or finely and shallowly wrinkled; older twigs with rough whitish flaking bark. Spines either furcate or bifurcate, the axis 3–12 mm. long, the first branches 5–12 mm., the second branches 6–16 mm. long. Leaves moderately coriaceous, drying discolorous, the upper surface dark grey or brown, the lower surface paler, green or brown; petiole c. 2 mm. long, glabrous or minutely puberulous. Lamina 1·5–3·5(5) × 0·7–2(3) cm., ovate to narrowly ovate or elliptic, rarely subcircular, with apex usually acute, rarely obtuse,

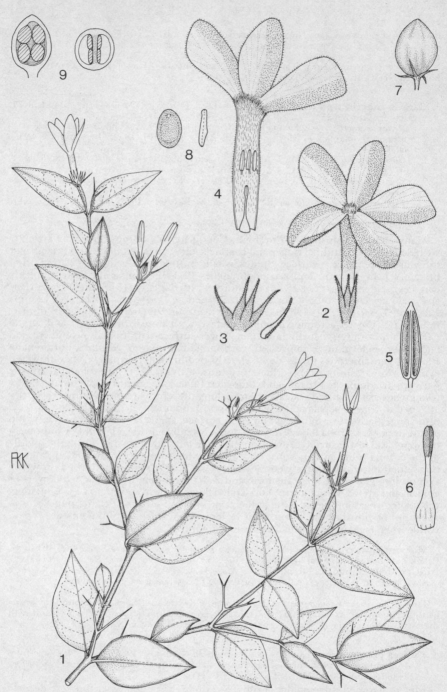

Tab. 90. CARISSA PRAETERMISSA. 1, habit (× ⅔), from *Mogg* 28448; 2, flower (× 2); 3, calyx, one sepal reversed (× ⁸⁄₃); 4, part of flower opened out (× ⁸⁄₃); 5, stamen (× 10), 2–5 from *Mendonça* 1974; 6, gynoecium (× 6), from *Gomes e Sousa* 1824; 7. fruit (× 1); 8, seed, dorsal and lateral views (× 2); 9, diagrammatic vertical and transverse sections through fruit showing the 8 seeds (hatched) borne on a median septum (stippled) (× 1), 7–9 from *Mendonça* 1824.

mucronate and base rounded to cuneate; upper surface glabrous, midrib impressed, lateral nerves faintly raised, very inconspicuous; lower surface glabrous or sparsely puberulous on midrib, midrib raised, lateral nerves invisible; margin revolute. Flowers pentamerous, solitary between two spines, shortly pedicellate. Calyx 4–5 mm. long, sepals subulate, dorsally puberulent. Corolla white; tube 11–16 mm. long, pubescent externally, pilose internally in the upper $\frac{2}{3}$ to $\frac{3}{4}$; lobes 9–21 mm. long, elliptic, overlapping to the left, pubescent below, glabrous or pubescent above. Stamens inserted at or just above the middle of the corolla tube; anthers subsessile, 1·5–2·5 mm. long. Gynoecium c. 4·5 mm. long, stigma not reaching to base of anthers; ovary c. 1 mm. long, ovoid, smooth, glabrous; style tapering, glabrous; stigma c. 1 mm. long, 2-lobed. Fruits c. 1·5 cm. long when ripe, subglobose, red, edible, 8-seeded. Seeds c. 5 mm. long, ovate in outline, discoid.

Mozambique. Z: Bajone, 3·2 km. from Murroa to Namuera, fl. 2.x.1949, *Barbosa & Carvalho* in *Barbosa* 4271 (K). GI: Inhambane, Panda, fl. 25.ii.1955, *E. M. & W.* 595 (LISC). Known only from Mozambique. In open woodland and forest, on sandy soil.

5. **Carissa bispinosa** (L.) Desf. ex Brenan in Mem. N.Y. Bot. Gard. **8**: 502 (1954). — Codd in Bothalia **7**: 450 (1961); in Fl. Southern Afr. **26**: 255 (1963). — Dale & Greenway, Kenya Trees & Shrubs: 44 (1961). — Compton, Fl. Swaziland: 439 (1976). — Kupicha in Bol. Soc. Brot. Sér 2, **53**, 1: 313 (1981). Type: Miller, Ic. **2**: t. 300 (1760).

Subsp. **bispinosa** — Kupicha, loc. cit. (1981).
  Arduina bispinosa L., Mant. Pl.: 52 (1767). Type as above.
  Lycium cordatum Miller, Gard. Dict., ed. 8: no. 10 (1768), *nom. illegit.* Type as above.
  Carissa arduina Lam., Encycl. Méth., Bot. **1**: 555 (1785), *nom. illegit.* Type as above.
  Carissa bispinosa (L.) Desf., Tabl. École. Bot.: 78 (1804), *nom. nud.*
  Carissa myrtoides Desf., Cat. Hort. Paris, ed. 3: 398 (1829). Type a cultivated plant.
  Arduina erythrocarpa Ecklon in S. Afr. Quart. Journ. **1**: 372 (1830). Type from S. Africa (Cape Prov.).
  Arduina acuminata E. Mey., Comm. Pl. Afr. Austr.: 191 (1837). Type from S. Africa (Cape Prov.).
  Carissa erythrocarpa (Ecklon) A. DC. in DC., Prodr. **8**: 335 (1844). Type as for *Arduina erythrocarpa.*
  Carissa acuminata (E. Mey.) A. DC., loc. cit. (1844). Type as for *Arduina acuminata.*
  Jasminonerium bispinosum (L.) Kuntze, Rev. Gen. Pl. **2**: 415 (1891). Type as for *Carissa bispinosa.*
  Arduina megaphylla Gand. in Bull. Soc. Bot. Fr. **65**: 59 (1918). Type from S. Africa (Cape Prov.).
  Carissa cordata (Miller) Fourc. in Trans. Roy. Soc. S. Afr. **21**: 82 (1934), *nom. illegit.* Type as for *Lycium cordatum.*
  Carissa dinteri Markgraf in Notizbl. Bot. Gart. Berl. **15**: 750 (1942). Type from Namibia.
  Carissa bispinosa var. *bispinosa* — Codd, loc. cit. (1961); op. cit.: 256, fig. 37, 1 (1963). — R. B. Drumm. in Kirkia **10**: 268 (1975).
  Carissa bispinosa var. *acuminata* (E. Mey.) Codd, tom. cit.: 451 (1961); op. cit.: 257 (1963) *excl.* fig. 37, 2. Type as for *Arduina acuminata.*

Sprawling shrub 0·6–3 m. tall. Young stems glabrous or minutely papillose, stout, bright yellowish or greyish green, with rugose surface; older stems similar in colour and texture. Spines numerous and conspicuous, as stout as the subtending stem, usually furcate, rarely simple or bifurcate; axis 21–38 mm. long, branches 15–46 mm. long. Leaves thickly coriaceous, glabrous, drying greyish or yellowish green, not or scarcely discolorous; petiole 2–3 mm. long, glabrous. Lamina 1·3–5·7 × 1–3·5 cm., ovate or subcircular, with apex obtuse, often mucronate and base truncate; upper surface with midrib very slightly channelled, other veins obscure; lower surface with midrib raised, other veins obscure; margin revolute. Inflorescence (1–)many-flowered, usually condensed, subsessile. Flowers strongly scented, corolla white inside, white or pink to greenish outside. Calyx 2·5–3 mm. long, segments lanceolate-subulate, dorsally glabrous, sparsely and minutely ciliate. Corolla tube 5–9 mm. long, widened slightly in upper half, externally glabrous or puberulent, internally glabrous or sparsely pubescent in upper half, and with a ring of hairs at the mouth. Corolla lobes ovate-acuminate, c. $\frac{1}{2}$ as long as tube, glabrous or puberulent on lower surface, glabrous above, overlapping to the left. Stamens inserted near the top of the corolla tube; anthers subsessile, 1·2–1·6 mm. long. Ovary c. 0·5 mm. long, subglobose, glabrous; style slender, glabrous; stigma at level of base of anthers or above. Fruit 1–1·3 cm. long, ovoid, red ripening black, edible, 1–2-seeded.

Botswana. N: Chobe National Park, st. 24.x.1969, *Mahundu* 54 (SRGH). SE: Gaborone, 995 m., st. 12.viii.1974, *Mott* 323 (K; SRGH). **Zimbabwe.** C: Gweru (Gwelo) Distr., 11 km. S. of Gweru, 1400 m., fl. 7.x.1966, *Biegel* 1346 (LMU; SRGH). W: Matobo Distr., Rhodes Matopos Estate, fl. & fr. 1.x.1951, *Plowes* 1267 (SRGH). S: Victoria Distr., Kyle Dam near Chembizi Gully, fr. 25.v.1971, *Mavi* 1242 (K; LISC; PRE; SRGH). **Mozambique.** GI: Inhambane Prov., Inharrime, Ponta Zavora, fl. 16.x.1957, *Barbosa & Lemos* in *Barbosa* 8080 (COI; K; LISC). M: Marracuene, Costa do Sol, fl. 10.viii.1959, *Barbosa & Lemos* in *Barbosa* 8659 (COI; K; LISC; PRE; SRGH).

Occurring also in S. Africa (western Transvaal, Natal and Cape Prov.) and Namibia and Kenya. On termitaria in *Brachystegia* woodland, in woodland fringes, and on maritime dunes.

The description of *C. bispinosa* subsp. *bispinosa* given above is based entirely on specimens from the F.Z. area (see note after subsp. *zambesiensis*).

Subsp. **zambesiensis** Kupicha in Bol. Soc. Brot., Sér. 2, **53**, 1: 321 (1980). Type: Zimbabwe, Mutare (Umtali) Distr., Nyamakwarara valley, fl. 2.xi.1967, *Mavi* 438 (K, holotype; LISC, PRE, SRGH, isotypes).

Carissa bispinosa var. *acuminata* sensu Codd in Bothalia **7**: 257 (1961), *quoad* fig. 37, 2 *excl. typ.* — sensu R. B. Drumm. in Kirkia **10**: 268 (1975).

Shrub 0·6–3 m. tall. Young stems glabrous or puberulous, with finely wrinkled surface, brownish-green, older stems similar. Spines sometimes absent, when present relatively few, inconspicuous, distinctly more slender than the subtending stem, usually furcate, rarely bifurcate; axis 2–13 mm. long, branches 3–12 mm. long. Leaves thinly coriaceous, glabrous, drying concolorous green; petiole 2–4 mm. long, glabrous. Lamina 2·5–11 × 1·1–4·5 cm., lanceolate, ovate or elliptic, the apex acute to acuminate, mucronate, the base cuneate to rounded. Inflorescence few–many-flowered, loose, shortly pedunculate. Flowers sweet-scented, corolla white inside, white to pinkish-green outside. Calyx 2–2·5 mm. long, segments lanceolate-subulate, minutely puberulous. Corolla tube 4·5–11 mm. long, ± straight-sided, glabrous externally, ± pubescent internally in upper half, with a ring of hairs at the mouth. Corolla lobes ovate-acuminate, ⅓ as long as to equalling corolla tube, overlapping to the left. Stamens inserted at the middle of the corolla tube; anthers subsessile, 0·8–1·5 mm. long. Ovary subglobose, glabrous; style slender, glabrous; stigma not reaching to base of anthers. Fruit 1·2–1·5 cm. long, narrowly ellipsoid, scarlet, edible, 1–2-seeded. Seeds compressed-ellipsoid, c. 1 cm. long.

Zimbabwe. E: Vumba Mts., 19 km. S. of Mutare (Umtali), fl. 18.xi.1960, *Angus* 2442 (FHO; K; LISC; SRGH). **Malawi.** C: Dedza Distr., Dedza Mt., 2010 m., fr. 18.v.1963, *Chapman* 2059 (SRGH). S: Mulanje Mt., near Thuchila Cottage, fr. 19.vii.1956, *Jackson* 2002 (BR; FHO; PRE). **Mozambique.** N: Ribáuè Mt., c. 1600 m., fr. 28.i.1964, *Torre & Paiva* 10315 (C; LISC; LMA; MO; SRGH; WAG). Z: Gúruè, near R. Malema, c. 1200 m., fl. 6.xi.1967, *Torre & Correia* 15949 (LISC; LMU). MS: Mocuta Mt., 800 m., fr. 6.vi.1971, *Müller & Gordon* 1817 (K; LISC; SRGH).

Also found in Swaziland and eastern Transvaal. In understorey of evergreen forest.

In the F.Z. area subsp. *bispinosa* and subsp. *zambesiensis* are completely distinct in vegetative and floral characters, habitat and distribution. In S. Africa, however, subsp. *bispinosa* is much more variable in vegetative characters and many specimens are indistinguishable from subsp. *zambesiensis* except by their flowers. The following specimen is the only example of an intermediate between the two subspecies known from the F.Z. area: Mozambique, M: Namaacha, Changalane, Estatuene, in gallery forest, fl. 10.xi.1967, *Gomes e Sousa & Balsinhas* 4999 (PRE). It has flowers with the anthers inserted at the middle of the corolla tube but with the stigma well above this level. Other intermediates have been seen from eastern Transvaal.

2. ACOKANTHERA G. Don

Acokanthera G. Don, Gen. Syst. **4**: 485 (1838). — Markgraf in Notizbl. Bot. Gart. Berl. **8**: 459 (1923). — Codd in Bothalia **7**: 448 (1961). — Kupicha in Kew Bull. **37**: 41 (1982).

Toxicophlaea Harv. in Hook., Journ. Bot. Lond. **1**: 24 (1842).

Trees and shrubs with poisonous sap, without spines or tendrils. Stipules absent. Flowers in dense axillary cymose fascicles. Calyx lobes imbricate, free to base. Corolla hypocrateriform; tube cylindrical, ± straight-sided, the inner surface usually pilose in the upper half and wrinkled below; lobes contorted, ovate, elliptic or subcircular, much shorter than the tube, overlapping to the left. Stamens inserted in the top of the corolla tube; anthers subsessile, without carina, with a tuft of hairs at apex. Ovary glabrous, bicarpellate, with 1 ovule per loculus; style filiform, glabrous, reaching to level of anthers; stigma minutely bifid, hairy, with a ring of papillae below. Fruit a globose to ellipsoid berry containing 1 or 2 seeds. Seeds plano-convex, with glabrous testa.

A genus of 5 species distributed from the Cape Prov. of S. Africa through east Africa to the Yemen.

Within the range of the genus, species of *Acokanthera* have had great importance as the chief source of arrow-poison.

1. Leaf-lamina almost always widest above the middle (obovate); lateral veins prominent on the upper leaf-surface, the proximal ones characteristically reaching the margin without joining their neighbours - - - - - - - - - 1. *oppositifolia*
   - Leaf-lamina of various shapes, seldom obovate; lateral veins not or slightly raised, inconspicuous, on the upper leaf-surface, each one looping to join its neighbour: - 2
2. Lamina 4–7·5 cm. long, less than twice as long as broad, usually scabrid (like sandpaper to the touch) - - - - - - - - - - - - - 2. *rotundata*
   - Lamina not as above, usually longer and always smooth to the touch - - - 3
3. Corolla tube c. 10 mm. long; leaf-lamina usually ovate - - - - 3. *laevigata*
   - Corolla tube 14–20 mm. long; leaf-lamina usually elliptic - - - 4. *oblongifolia*

1. **Acokanthera oppositifolia** (Lam.) Codd in Bothalia **7**: 448 (1961); in Fl. Southern Afr. **26**: 247, fig. 36, 1 (1963). — Fanshawe, Check List Woody Pl. Zambia: 2 (1973). — R. B. Drumm. in Kirkia **10**: 268 (1975). Kupicha in Kew Bull. **37**: 53 (1982). TAB. **91**. Type from Africa (origin uncertain).
   Cestrum venenatum Thunb., Prodr.: 36 (1794), *nom. illegit.* Type from S. Africa.
   Cestrum oppositifolium Lam., Tabl. Encycl. Méth., Bot. tom. 1 vol. **1** part 2: t. 112, fig. 2 (1792); tom. 2 vol. **3** part 1: 5 (1794). Type as for *Acokanthera oppositifolia*.
   Acokanthera lamarckii G. Don, Gen. Syst. **4**: 485 (1838), *nom. superfl. illegit.* Type as for *Acokanthera oppositifolia.*
   Toxicophlaea thunbergii Harv. in Hook., Journ. Bot. Lond. **1**: 24 (1842), *nom. nov. pro Cestrum venenatum* Thunb. *non* Burm. f.
   Toxicophlaea cestroides A. DC. in DC., Prodr. **8**: 336 (1844), *nom. superfl. illegit.* Type as for *Acokanthera oppositifolia.*
   Toxicophlaea thunbergii var. *scabra* Sond. in Linnaea **23**: 79 (1850). Lectotype from S. Africa (Natal).
   Pleiocarpa hockii De Wild. in Fedde, Repert. **13**: 109 (1914). Lectotype from Zaire (Shaba).
   Acokanthera longiflora Stapf in Kew Bull. **1922**: 28 (1922). — Markgraf in Notizbl. Bot. Gart. Berl. **8**: 468, fig. 6 (1923). — Dale & Greenway, Kenya Trees & Shrubs: 43 (1961). Lectotype from Tanzania.
   Acokanthera venenata G. Don var. *scabra* (Sond.) Markgraf, tom. cit.: 470 (1923). Lectotype as for *Toxicophlaea thunbergii* var. *scabra.*
   Carissa acokanthera Pichon in Mém. Mus. Hist. Nat. Paris, N.S. **24**: 132 (1948), *nom. superfl. illegit.* — F. White, F.F.N.R.: 347 (1962). Type as for *Cestrum venenatum* Thunb.
   Carissa oppositifolia (Lam.) Pichon in Bull. Jard. Bot. Brux. **22**: 109 (1952). Type as for *Acokanthera oppositifolia.*
   Acokanthera rhodesica Merxm. in Mitt. Bot. Staatss. München **1**: 201 (1953). Type from Zimbabwe, Rusape Distr., fl. & fr. 15.x.1952, *Dehn* 36/52 (SRGH, isotype).
   Garcinia sciura Spirlet in Bull. Jard. Bot. Brux. **29**: 327 (1959). Type from Zaire (Shaba).
   Carissa longiflora (Stapf) Lawrence in Baileya **7**: 90 (1959). Lectotype as for *Acokanthera longiflora.*
   Acokanthera venenata auct. mult., *non* G. Don. (1838).

A much-branched evergreen shrub, sometimes scrambling, or small tree, 1–6 m. high; bark brown, deeply fissured; slash cream turning olive green. Young branches with reddish tinge, glabrous, conspicuously angled and ribbed. Leaves coriaceous, glabrous, smooth or occasionally slightly scabrid; petiole 2–6(7) mm. long; lamina 4·6–10·6 × 1·8–6·7 cm., usually obovate, occasionally elliptic, the apex acute, cuspidate or obtuse, with hard mucro, the base cuneate or rounded; upper surface glossy, with midrib impressed and lateral nerves strongly raised; lower surface mat, all nerves raised, the proximal lateral nerves characteristically reaching the margin without looping to join neighbouring nerves; midrib of lower surface and petiole wrinkled. Inflorescences very plentiful, dense, contracted, many-flowered axillary cymes; flowers fragrant, with pink or reddish corolla tube and white lobes. Calyx 2–3 mm. long, lobes ovate to lanceolate, weakly imbricate, dorsally pilose and ciliate or glabrous. Corolla tube 9–13·5 mm. long, pubescent or hispid on external surface or rarely glabrous, pilose within above the middle and wrinkled below; lobes ovate-cuspidate, 2–4·6 mm. long, pubescent or glabrous on either side, usually ciliate. Anthers 1·1–1·4 mm. long, visible in the mouth of the corolla tube at anthesis. Ovary 0·6–1 mm. long, ellipsoid, smooth. Fruit an ellipsoid berry 12–20 mm. long (dry), purple when ripe; seeds 6·5–10 mm. long.

Zambia. W: Ndola Distr., near Mwekera Rest House near Nkana, fl. 27.ix.1947, *Brenan & Greenway* in *Brenan* 7982 (BM; BR; FHO; K; PRE). **Zimbabwe.** N: Urungwe Reserve, fr. 27.ix.1952, *Phelps* 28 (LISC; K; SRGH). W: Bulawayo, fl. vi.1898, *Rand* 572 (BM). C: Enterprise Distr., by Umwindsi R., fl. 14.vii.1946, *Wild* 1175 (K; PRE; SRGH). E: Mutare (Umtali) Distr., Commonage, fr. 12.i.1949, *Chase* 1580 (BM; COI; LISC; K; SRGH). S: Masuingo (Fort Victoria, Nyanda) near Great Zimbabwe (Zimbabwe ruins), fr. 7.xii.1961, *Leach* 11301 (K; LISC; PRE; SRGH). **Malawi.** N: Ekwendeni, near Kasitu R., fl. 13.vi.1954, *Jackson* 1338 (BR; FHO; K; MAL). **Mozambique.** GI: Gaza, Muchopes, 25 km. from Manjacaze, road by Chidenguele, fr. 17.iii.1948, *Torre* 7508 (C;LISC;LMA;MO;WAG). M: Maputo (Lourenço Marques), fl. 11.viii. 1920, *Borle* 545 (K; PRE; SRGH).

Widely distributed in eastern Africa from the Cape Prov. to Zaire (Shaba), Tanzania and Kenya. In riverine vegetation in shade of taller trees, and also often on termitaria.

Fanshawe (loc. cit.) records *A. oppositifolia* also from Zambia (B and E), but I have seen no specimens from these areas. Specimens of *A. oppositifolia* from Mozambique often have leaves rather different from those in the rest of the species' range: the lamina tends to be elliptic rather than obovate, and to have looping lower lateral veins. In the vegetative state, such specimens are difficult to distinguish from *A. oblongifolia* from the same area; the only differences remaining between the leaves of the two species are the longer petioles and less prominent venation of *A. oblongifolia*. However, flowering and fruiting material of these species is easily separated, *A. oblongifolia* having distinctly longer flowers and larger fruits than *A. oppositifolia*.

2. **Acokanthera rotundata** (Codd) Kupicha in Kew Bull. **37**: 60 (1982). Type from S. Africa (Transvaal).
　　Acokanthera schimperi (A. DC.) Benth. var. *rotundata* Codd in Bothalia **7**: 449 (1961); in Fl. Southern Afr. **26**: 249, fig. 36, 3 (1963). — R. B. Drumm. in Kirkia **10**: 268 (1975). Type as above.

Shrub or small much-branched tree up to 5 m. high and 5 m. in diameter, with rough bark. Young branches pubescent or less often glabrous, conspicuously angled and ribbed. Leaves very coriaceous, often drying somewhat concave, slightly to strongly papillose-scabrid, especially above; petiole 3–7 mm. long, pubescent on upper surface; lamina 3·8–7·5 × 2·5–4·8 cm., broadly elliptic to subcircular, the apex obtuse with blunt or sharp mucro, the base rounded; upper surface glossy, with midrib level and lateral veins faintly raised or impressed, glabrous except for a line of hairs sometimes continuing from the petiole along the midrib; lower surface mat, midrib raised, laterals faintly so; lateral veins looped to join their neighbours. Inflorescences dense contracted many-flowered axillary cymes; flowers sweet-smelling, with pink or red corolla tube and white lobes. Calyx 2–2·5 mm. long, lobes ovate-acute with often slightly recurved tip, strongly imbricate, dorsally pubescent and ciliate. Corolla tube 9–12 mm. long, glabrous to minutely pubescent on external surface, long-pilose above and wrinkled below, on inner surface; corolla lobes subcircular with apex rounded or cuspidate, 3–5·5 mm. long, ciliate or not, pubescent or glabrous. Stamens inserted at the top of the corolla tube so that the anther tips are just visible at anthesis; anthers 1–1·3 mm. long. Ovary 0·6–0·7 mm. long, conical. Ripe fruit up to 2 cm. in diameter, subglobose, red to purple, 1–2-seeded. Seeds up to 1 cm. long.

Zimbabwe. N: Chipuriro (Sipolilo) Distr., Nyarasuswe, st. x.1959, *Orpen* s.n. (SRGH). W: Matobo Distr., Farm Besna Kobila, 1490 m., fl. buds ii.1959, *Miller* 5751 (FHO; K; PRE; SRGH). S: Bikita Distr., Magahane Hill, 1.viii.1978, *Hall* 58 (SRGH).

Also found in Swaziland, N. and E. Transvaal and N. Natal. On cliffs, among rocks and in open grassland.

3. **Acokanthera laevigata** Kupicha in Kew Bull. **37**: 58 (1982). Type from Tanzania.

Shrub (in Malawi) or tree up to 12 m. high, with white wood. Young branches glabrous, compressed with finely striate surface or sometimes strongly ribbed. Leaves coriaceous, glabrous, smooth, drying ± concolorous brown or green; petiole 3–10 mm. long; lamina 5·5–12·25 × 3·25–6·7 cm., ovate or ovate-elliptic, with apex obtuse or acute and base rounded or cuneate; upper surface glossy, with midrib impressed and lateral veins level or faintly raised or impressed, inconspicuous; lower

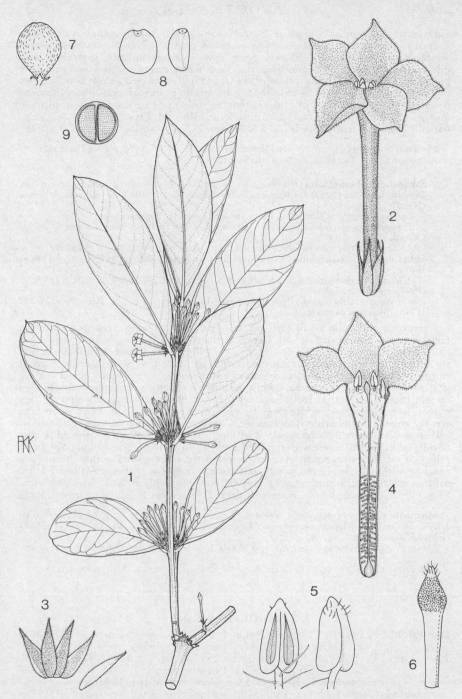

Tab. 91. ACOKANTHERA OPPOSITIFOLIA. 1, habit (×⅔); 2, flower (×4); 3, calyx, one sepal reversed (×4); 4, flower opened out (×4); 5, stamen, ventral and dorsal views (×20); 6, apex of gynoecium (×20); 7, fruit (×1); 8, seed, dorsal and lateral views (×1) 9, diagrammatic transection through fruit showing the two seeds (stippled) separated by median septum. 1–6 from *Fanshawe* 2395; 7–9 from *Biegel* 2462.

surface mat, with midrib raised and lateral veins level or faintly raised; lateral veins looped to join their neighbours. Inflorescences contracted axillary cymes; flowers white, fragrant. Calyx 2·5–3·5 mm. long, lobes ovate-acuminate, sparsely pilose, ciliate. Corolla tube c. 10 mm. long, the outer surface very sparsely and minutely pubescent, the inner surface pilose in the upper half, wrinkled below; corolla lobes c. 3 mm. long, broadly ovate-cuspidate, glabrous on both surfaces, ciliate. Stamens inserted at the top of the corolla tube so that the anther tips are just visible at anthesis; anthers c. 1·3 mm. long. Ovary c. 1 mm. long, ellipsoid. Fruit 2·5–3·2 × 2–2·5 cm. (dried), 2-seeded, ovoid. Seeds c. 8·5 × 7·5 mm.

Malawi. N: Rumphi Distr., Nyika Plateau, 2160 m., fr. 11.viii.1976, *Pawek* 11818 (K). Also occurring in Tanzania. In *Juniperus* forest.

4. **Acokanthera oblongifolia** (Hochst.) Codd in Bothalia **7**: 449 (1961); in Fl. Southern Afr. **26**: 246, fig. 36, 2 (1963). — Kupicha in Kew Bull. **37**: 62 (1982). Type from S. Africa (Natal).
  Carissa oblongifolia Hochst. in Flora **1844**: 827 (1844). Type as above.
  Toxicophlaea spectabilis Sond. in Linnaea **23**: 79 (1850). Syntypes from S. Africa (Natal).
  Acokanthera spectabilis (Sond.) Hook. f., Bot. Mag.: t. 6359 (1878). — Stapf in F. C. **4**: 501 (1907). — Markgraf in Notizbl. Bot. Gart. Berlin **8**: 470, fig. 9 (1923). Syntypes as above.
  Acokanthera venenata var. *spectabilis* (Sond.) Sim., For. Fl. Cape Col.: 270, t. 154 fig. 2 (1907). Syntypes as above.
  Carissa spectabilis (Sond.) Pichon in Mém. Mus. Nation. Hist. Nat. Paris, N.S. **24**: 132 (1948). Syntypes as above.

Evergreen shrub or small tree up to 6 m. high. Young branches glabrous, conspicuously angled and ribbed. Leaves coriaceous, glabrous, smooth; petiole 4–9 mm. long; lamina 6–8·4 × 1·5–4·5 cm., ± elliptic, the apex obtuse to acute, mucronate, the base cuneate or rounded; upper surface glossy, with midrib shallowly impressed and lateral veins slightly raised but usually inconspicuous; lower surface mat, midrib and lateral veins raised but the latter inconspicuous; lateral veins looped to join their neighbours. Inflorescences dense contracted many-flowered axillary cymes; flowers fragrant, white tinged pink. Calyx c. 3 mm. long, lobes lanceolate, weakly imbricate, puberulous and ciliate. Corolla tube 14–20 mm. long, glabrous or pubescent on external surface, pilose within in upper half, faintly wrinkled below; corolla lobes broadly ovate with rounded apex, 3–7 mm. long, glabrous to pubescent, ciliate or not. Stamens inserted near the top of the corolla tube so that the anthers reach to within 1 mm. of the mouth and are not visible at anthesis; anthers 1·5–1·7 mm. long. Ovary c. 1 mm. long, cylindrical, longitudinally ribbed. Fruit 2–2·5 cm. long, ellipsoid or subglobose, purplish-black, 2(1)-seeded. Seeds up to 1·5 cm. long.

Mozambique. GI: Inhambane, Praia de Zavora, fr. 27.ii.1955, *E.M. & W.* 688 (BM; LISC; SRGH). M: Maputo (Lourenço Marques), Costa do Sol, fr. 20.vi.1944, *Torre* 6641 (BM; LISC).
Also in S. Africa (eastern Cape Prov. and Natal). In sclerophyllous maquis on littoral dunes.

Leaf-lamina lengths of up to 12·25 cm. have been recorded from S. African specimens.

3. LANDOLPHIA P. Beauv.

Landolphia P. Beauv., Fl. Oware Bénin **1**: 54 (1806), *nom. conserv.* — Pichon in Mém. Inst. Fr. Afr. Noire **35**: 40 (1953).
Vahea Lam., Tabl. Encycl. Méth., Bot. tom. 1 vol. **1**, part 2: t. 169 (1792); tom. 2 vol. **5**, part 2: 292 (1819), *nom. rejic.*
Carpodinus R. Br. ex G. Don, Gen. Syst. **4**: 101 (1838).
Jasminochyla (Stapf) Pichon in Bull. Mus. Nation. Hist. Nat., Sér. 2, **20**: 550 (1949).

Lianes, shrubs and suffrutices, tendrillous or rarely lacking tendrils (not in F.Z. area). Latex present in all parts, often abundant. Spines absent. Stipules absent. Cymes terminal and/or axillary, often grouped into terminal panicles, peduncle usually short. Calyx lobes imbricate, free or shortly united at the base. Corolla hypocrateriform, glabrous or variously hairy; tube ± cylindrical, slightly enlarged at level of anthers; lobes broadly elliptic to narrowly linear, contorted, overlapping to the left. Stamens inserted at various levels in the corolla tube; filaments very short; anthers dorsi- or basi-fixed, introrse, without carina. Ovary glabrous or hairy, uni- or bilocular, placentation axile or parietal with very prominent placentae, ovules ∞. Style glabrous or pilose; clavuncle always positioned at level of base of anthers, of various shapes; stigma bifid but with the arms connivent. Fruit a many-seeded berry with or without sclerified pericarp; pulp sweet and acid. Seeds endospermous, testa firmly united with the placental pulp.

A genus of 55 species occurring in Africa, Madagascar and the Mascarenes.

The tendrils of *Landolphia* and other genera of the Landolphiinae (*Ancylobothrys, Aphanostylis, Dictyophleba, Saba*) are homologous with an inflorescence, and are always terminal.

1. Inflorescences terminal and sometimes also axillary - - - - - - 2
 - Inflorescences all axillary - - - - - - - - - - - 7
2. Corolla tube 15 mm. long or more - - - - - - - - - - 3
 - Corolla tube 9 mm. long or less - - - - - - - - - - 4
3. Flowers in lax 1–4-flowered cymes with axes glabrous or hispidulous; corolla glabrous, lobes narrowly linear; style glabrous, stamens inserted near top of corolla tube - 2. *camptoloba*
 - Flowers in dense many-flowered capitate panicles with axes ferrugineous-pilose; corolla pubescent, lobes elliptic, 3·5–8 mm. broad; style densely pilose in the lower ⅔; stamens inserted in the middle of corolla tube - - - - - - 6. *eminiana*
4. Corolla tube 7·5–9 mm. long, with stamens inserted in the middle; exterior surface of tube glabrous or sparsely pilose, not puberulent; upper leaf surface glabrous, with lateral veins and tertiary reticulation impressed, giving a finely cracked appearance - 1. *buchananii*
 - Corolla tube 6·5 mm. long or less, with stamens inserted near the apex; exterior surface of tube puberulent; upper leaf surface glabrous or pubescent, especially on midrib, with lateral veins and tertiary reticulation raised or sometimes level, but not giving a cracked appearance - - - - - - - - - - - - - - - 5
5. Leaf lamina 7·5–11·6 × 3–4·4 cm.; upper surface glabrous; corolla tube 6–6·5 mm. long
 5. *owariensis*
 - Leaf lamina usually much smaller, up to 9·2 × 3·2 cm.; upper surface almost invariably with some indumentum, at least on midrib; corolla tube 3·5–4·5 mm. long - - - 6
6. Leaves relatively small (up to 5·2 × 2·6 cm.), yellowish-green, characteristically closely and neatly arranged; flowers white or cream-coloured; sepals glistening reddish-brown; exterior surface of corolla lobes puberulent on overlapping surfaces - - - 4. *parvifolia*
 - Leaves often larger than above, brownish- or greyish-green, more loosely and untidily arranged; flowers with corolla yellow on outer surface, lobes white above; sepals mat, greyish-brown; corolla lobes glabrous - - - - - - - - - 3. *kirkii*
7. Suffrutex or low shrub up to 1 m. high; leaves with lamina 4·5–8·5 × 0·8–4 cm.; flower attaining 19 mm. long (petals unopened), corolla lobes longer than tube - 8. *cuneifolia*
 - Liane, climbing much higher than 1 m.; leaves with lamina 5·5–15 × 3–7·5 cm.; flower attaining 10 mm. long (petals unopened), corolla lobes shorter than tube - 7. *rufescens*

1. **Landolphia buchananii** (Hallier f.) Stapf in F.T.A. **4**, 1: 35 (1902). — Pichon in Mém. Inst. Fr. Afr. Noire **35**: 48, t. 1 fig. 4–9 map 1 (1953). — F. White, F.F.N.R.: 349 (1962). — Fanshawe, Check-list Woody Pl. Zambia: 25 (1973). — R. B. Drumm. in Kirkia **10**: 268 (1975). Lectotype: Malawi, Shire Highlands, 1891, *Buchanan* 220 p.p.* (BM, isotype; K, holotype).

Clitandra? buchananii Hallier f. in Jahrb. Hamb. Wiss. Anst. **17**, Beih. 3: 118 (1900, October or later). Type as above.

Clitandra kilimandjarica Warb. in Tropenpfl. **4**: 614 (December 1900). Type from Tanzania.

Landolphia kilimandjarica (Warb.) Stapf, tom. cit.: 34 (1902). Type as above.

Landolphia cameronis Stapf, tom. cit.: 35 (1902). Syntypes from Malawi, *Cameron* 1 (K) and 11 (K).

Landolphia ugandensis Stapf, tom. cit.: 589 (1904). Type from Uganda.

Landolphia swynnertonii S. Moore in Journ. Linn. Soc., Bot. **40**: 136 (1911). Type: Zimbabwe, Chirinda Forest, 1128–1219 m., fl. xi, *Swynnerton* 82 (BM, holotype; K, SRGH, isotypes).

Landolphia rogersii Stapf in Kew Bull. **1912**: 360 (1912). Type from Zaire.

Clitandra semlikiensis Robyns & Boutique in Bull. Jard. Bot. Brux. **18**: 259 (1947). Type from Zaire.

Jasminochyla ugandensis (Stapf) Pichon in Bull. Mus. Nat. Hist. Nat., Sér. 2, **20**: 551 (1948). Type as for *Landolphia ugandensis*.

* At least three different species of Apocynaceae are known within the collection *Buchanan* 220: the lectotype of *L. buchananii*, *L. kirkii* (2 sheets at K) and *Saba comorensis* (3 sheets at K, including both varieties).

Scandent shrub or liane attaining up to 40 m. with support from trees. Stems deeply fluted "like a bundle of parallel ropes"; bark brown with grey spots. Young stems glabrous or hispidulous. Leaves membranous, drying brownish green, usually not strongly discolorous; petiole 3–8 mm. long, glabrous or pubescent, often rugose, deeply channelled and often folded into a tube; lamina (4)5–12(14) × (1·5)1·7–5 cm., ovate- or elliptic-oblong, the apex cuspidate-attenuate into a short round-tipped acumen, the base acute to rounded. Upper leaf surface mat or somewhat glossy, completely glabrous; midrib rather wide, scarcely channelled; secondary and tertiary veins impressed so that the surface bears a reticulum of shallow cracks. Lower surface completely glabrous or with pubescent midrib; midrib prominent, lateral nerves sometimes raised, tertiary reticulation level with surface, easily seen with a hand-lens. Tendrils common. Inflorescences terminal and axillary few- to several-flowered cymes or racemes, forming dense clusters or frequently elongating and becoming tendrillous; axes glabrous or pubescent. Flowers subtended by 2 sepal-like bracteoles, strongly jasmine-scented. Calyx of 5 strongly imbricate ovate sepals 1·5–2 mm. long, united at the base, glabrous with minutely ciliate margins, c. $\frac{1}{5}$–$\frac{1}{4}$ as long as corolla tube. Corolla white, the tube and reverse of lobes tinged pink or yellow; corolla tube 7·5–9 mm. long, slightly widened in the middle at the level of the anthers, outer surface glabrous or sparsely hairy; corolla lobes narrowly elliptic, ± as long as tube, glabrous. Stamens inserted at the middle of the corolla tube; anthers 1·5–1·75 mm. long. Ovary 1 mm. high, conical, glabrous; style, clavuncle and stigma 1·5–2 mm. long. Fruits up to 10 cm. in diameter, globose. Seeds 8–17 mm. long.

Zambia. N: Mbala (Abercorn) Distr., Vomo Gap, Fwambo side, 1500 m., fl. buds 27.ix.1960, *Richards* 13273 (K; SRGH). W: 16 km. W. of Kakoma on Zaire border, 97 km. NE. of Mwinilunga, fl. 30.ix.1952, *Holmes* 932 (FHO; K; SRGH). E: Lundazi Distr., upper slopes of Kangampande Mt., Nyika Plateau, 2134 m., st. 7.v.1952, *White* 2769 (BR; FHO; K). **Zimbabwe.** E: Inyanga Distr., Mtarazi National Park, 1200 m., fl. 19.x.1969, *Müller & Biegel* 1285 (COI; FHO; K; LISC; PRE; SRGH). **Malawi.** N: Wilindi (Walindi) Forest, Misuku Hills, 2000 m., fl.12.xi.1958, *Robson* 588 (BM; K; LISC; PRE; SRGH). C: Ntchisi forest, fr. 14.iii.1967, *Salubeni* 616 (K; LISC; MAL; PRE; SRGH). S: Soche Mt. near Blantyre, 1372 m., fl. viii.1944, *Benson* 485 (K; PRE). **Mozambique.** N: Malema, 40 km. from Entre-Rios on road to Ribáuè, Murripa Mt., c. 1100 m., fr. 15.xii.1967, *Torre & Correia* 16529 (LISC; LMU). Z: Milange, Tumbine Mt., near Vila Masseti, c. 1000 m., fr. 18.i.1966, *Correia* 445 (C; LISC; LMA; MO; SRGH; WAG). T: Mt. Zóbuè, fl. 3.x.1942, *Mendonça* 583 (BR; COI; EA; LISC; LMA; M; PRE). MS: Manica, Báruè, Choa Mt., 17 km. from Catandica (Vila Gouveia), c. 1500 m., fl. 13.xii.1965, *Torre & Correia* 13610 (K; LISC).

Also occurring in Zaire, Angola, Uganda, Rwanda, Kenya and Tanzania. In relict montane evergreen forest and riverine forest.

The collection *Brass* 16965 from central Malawi (K; PRE; SRGH) describes an unusual habit for this species: "on rocks in *Brachystegia* woodland".

2. **Landolphia camptoloba** (K. Schum.) Pichon in Mém. Inst. Fr. Afr. Noire **35**: 78, t. 1 fig. 11–14, map 4 (1953). — F. White, F.F.N.R.: 349 (1962). — Fanshawe, Check List Woody Pl. Zambia: 25 (1973). Type from Zaire.
    Carpodinus camptoloba K. Schum. in Engl. & Prantl, Nat. Pflanzenfam. **4**, 2: 132 (1895). Type as above.
    Carpodinus gracilis Stapf in Kew Bull. **1898**: 303 (1898). Syntypes from Zaire.
    Clitandra gracilis (Stapf) Hallier f. in Jahrb. Hamb. Wiss. Anst. **17**, Beih. 3: 117 (1900). Type as above.
    Carpodinus leucantha K. Schum. in Warb., Kunene-Sambesi Exped. Baum: 338 (1903). Type from Angola.

Rhizomatous scrambling shrub or climber attaining 3 m. Young shoots glabrous or hispidulous. Leaves membranous or thinly coriaceous, not strongly discolorous; petiole 2–5 mm. long, glabrous or hispidulous; lamina 4·5–6·8 × 1·4–2·7 mm., ovate, oblong or narrowly elliptic, tapering at apex to a long distinct acumen, base rounded to acute. Upper surface glossy, glabrous, midrib raised or level, secondary and tertiary veins faintly raised or impressed to give a finely "cracked" appearance; lower surface completely glabrous or with the midrib hispidulous, a similar indumentum often found also on the lamina near its base, midrib raised, lateral and tertiary nerves raised or impressed. Tendrils common. Flowers white, scented, in lax terminal 1–4-flowered cymes with axes glabrous or hispidulous. Calyx 2–3 mm. long, lobes united at base, ovate-acuminate, ciliate, dorsally glabrous or puberulent. Corolla tube

15–18 mm. long, slender, swollen near apex at level of stamens, glabrous, not constricted at the mouth; corolla lobes ± equalling tube, narrowly linear, glabrous. Anthers c. 2 mm. long, subsessile, inserted near the top of the corolla tube so that they reach to within 1 mm. of its mouth. Gynoecium glabrous. Ovary narrowly conical, c. 2 mm. long, passing imperceptibly into the long style. Stigma reaching to base of anthers. Fruit globose, c. 2 cm. in diameter, yellow or orange with spots when ripe, edible. Seeds 6–7·5 mm. long.

Zambia. B: 44 km. from Kaoma (Mankoya) to Mongu, fl. 8.xi.1959, *Drummond & Cookson* 6241 (K; LISC; PRE; SRGH). N: Mbala (Abercorn) Distr., new road to Iyendwe valley from Kambole, 1500 m., st. 31.i.1959, *Richards* 10800 (K). S: Machili, fl. 14.xi.1960, *Fanshawe* 5891 (FHO; K).

Occurring widely in Zaire and Angola. In *Brachystegia* and *Cryptosepalum* woodland, usually on Kalahari Sands.

3. **Landolphia kirkii** Dyer ex Hook. f. in Kew Gard. Rep. **1880**: 39 (1881); in Hook. f., Ic. Pl. t. 2755 (1903). — Dewèvre in Ann. Soc. Sci. Brux. **19**: 138 (1895). — Stapf in F.T.A. **4**, 1: 55 (1902). — Pichon in Mém. Inst. Fr. Afr. Noire **35**: 88, map 4 (1953). — F. White, F.F.N.R.: 349, fig. 62 J, K (1962). — Codd in Fl. Southern Afr. **26**: 259 (1963). — Fanshawe, Check List Woody Pl. Zambia: 25 (1973).   R. B. Drumm. in Kirkia **10**: 268 (1975). Described from the mouth of the Zambezi (Mozambique); type not traced.

Vahea kirkii (Dyer ex Hook. f.) Sadeb. in Jahrb. Hamb. Wiss. Anst. **9**, 1: 226 (1891). Based on *Landolphia kirkii*.

Landolphia kirkii var. *delagoensis* Dewèvre, tom. cit.: 140 (1895). Type: Mozambique, Delagoa Bay, *Junod* s.n. (not seen).

Landolphia delagoensis (Dewèvre) Pierre in Bull. Soc. Linn. Paris, N.S. **1**: 15 (1898). Type as above.

Landolphia polyantha K. Schum. in Engl., Bot. Jahrb. **28**: 452 (1900). Type from Tanzania.

Landolphia dondeensis Busse in Tropenpfl. **5**: 406 cum fig. (1901). Type from Tanzania.

Landolphia kirkii var. *dondeensis* (Busse) Stapf in F.T.A. **4**, 1: 56 (1902). Type as above.

Straggling shrub or liane attaining 18(30) m., with rough dark bark. Young stems brownish- or ferrugineous-pubescent, later glabrescent. Leaves membranous to subcoriaceous, usually drying discolorous with the upper surface much darker than the lower; petiole 2–7 mm. long, pubescent; lamina 2·4–9·2 × 1–3·2 cm., oblong to narrowly ovate or rarely oblong-obovate, the apex attenuate or cuspidate-attenuate into a short to long acumen, or acumen ± absent, the base rounded to cuneate. Upper leaf surface fairly glossy, almost always puberulent when young, glabrescent, or less often glabrous except for the pubescent midrib, very rarely lamina completely glabrous, midrib channelled, lateral and tertiary nerves more or less raised, never impressed; lower surface crispate-pubescent or glabrous except for the appressed- or spreading-pubescent midrib, midrib prominent, other nerves raised or level with surface, tertiary reticulation easily seen with a hand-lens. Tendrils common. Inflorescences terminal, many-flowered paniculate cymes forming dense clusters or sometimes the branches elongating and becoming tendrillous; axes densely rufous-tomentose or -pubescent. Flowers scented, each subtended by two sepal-like bracteoles and the inflorescence branches by bracts. Calyx c. 2 mm. long, sepals almost free, oblong, greyish-brown, not glossy, ciliate, dorsally glabrous or puberulent, with hispid midrib. Corolla white, cream-coloured or yellow, darker on the tube and reverse of petals; corolla tube 3·5–4 mm. long, clavate, the wall thickened in the upper half and the mouth very restricted; exterior surface puberulous especially above the calyx; corolla lobes narrowly elliptic, ± equalling the tube, glabrous. Stamens inserted at or just above the middle of the corolla tube; anthers subsessile, c. 0·75 mm. long, reaching to base of constricted part of corolla tube. Ovary 0·75–1 mm. long, glabrous; style, clavuncle and stigma 1–1·5 mm. long. Fruit globose or pyriform, attaining 15 cm. in diameter, green with white flesh, edible. Seeds 8·5–10 mm. long.

Zambia. N: Mbala (Abercorn) Distr., road to Mpulungu below Chizunga Scarp, 1372 m., fl. 7.x.1936, *Burtt* 6346 (BM; BR; K; FHO; PRE). W: Mansa (Fort Rosebery), Samfya Mission near L. Bangweulu, fl. 19.viii.1952, *Angus* 241 (BM; BR; FHO; K). C: Kabwe (Broken Hill) Distr., Broken Hill Forest Reserve, fl. 21.ix.1947, *Brenan & Trapnell* in *Brenan* 7903 (FHO; K). **Zimbabwe.** E: Mutare (Umtali) Distr., S. side of Murahwa's Hill, West Commonage, 1115 m., fl. 14.x.1962, *Chase* 7855 (BR; K; LISC; PRE; SRGH). **Malawi.** C: Dedza Distr.,

Namikokwe R., Mua Reserve, fl. 13.x.1965, *Banda* 693 (MAL; SRGH). S: Zomba Plateau, fl. ix.1895, *Whyte & McClounie* 3 (K). **Mozambique.** N: Ribáuè, Mepáluè Mt., c. 1400 m., fl. & fr. 5.xii.1967, *Torre & Correia* 16371 (C; LISC; LMU; SRGH; WAG). Z: Mathilde, fl. 9.x.1904, *Le Testu* 464 (BM; BR; K; LISC). MS: Chimoio, Vandúzi, fl. 19.x.1944, *Mendonça* 2511 (COI; LISC; LMA; MO; PRE). GI: Inhambane Prov., Massinga, fl. x.1936, *Gomes e Sousa* 1902 (BR; COI; K). M: Matutuine (Bela Vista), Tinonganine to Friere, Liguati forest, fr. 8.viii.1957, *Barbosa & Lemos* in *Barbosa* 7784 (COI; K; LISC).

Also in Zaire, Tanzania, Kenya and S. Africa (Natal). In *Brachystegia* woodland, deciduous forest, riverine forest and evergreen forest.

4. **Landolphia parvifolia** K. Schum. in Engl., Bot. Jahrb. **15**: 409 (1893). — Stapf in F.TA. **4**, 1: 57 (1902). — Pichon in Mém. Inst. Fr. Afr. Noire **35**: 94, map 5 (1953). — F. White, F.F.N.R.: 349 (1962). — Fanshawe, Check List Woody Pl. Zambia: 25 (1973). Type from Angola.

> *Pacouria parvifolia* (K. Schum.) Hiern, Cat. Afr. Pl. Welw. **1**: 663 (1898). Type as above.
> *Landolphia kirkii* var. *parvifolia* (K. Schum.) Hallier f. in Jahrb. Hamb. Wiss. Anst. **17**, Beih. 3: 74 (1900). Type as above.

Many-stemmed shrub or small tree up to 4 m. tall, with scandent branches climbing when supported. Young shoots densely brownish or ferrugineous hispid-pubescent, indumentum persisting until bark formation. Leaves characteristically neatly and closely arranged, orange-tinged when dry, yellowish-green, not strongly discolorous, coriaceous in texture; petiole 2–4 mm. long, hispid; lamina 2·6–5·2 × 1–2·6 cm., oblong, narrowly elliptic-ovate or elliptic-obovate, the apex cuspidate-attenuate into a short round-tipped acumen, the base cuneate to subcordate. Upper leaf surface glossy, usually glabrous except for the pubescent midrib and leaf-margins, occasionally the leaf margins also glabrous, occasionally the whole lamina puberulent; venation conspicuous, characteristically regular, the midrib channelled, the other nerves raised; lower leaf surface completely glabrous or with pubescent midrib, venation less conspicuous than above, midrib prominent, other nerves raised. Tendrils uncommon. Inflorescences many-flowered paniculate cymes borne terminally on the main stems and on lateral branches, and rarely also in leaf axils; panicles usually forming compact clusters but very occasionally lengthening greatly and becoming tendrillous; axes tomentose to pubescent. Flowers subtended by two sepal-like bracteoles, white or creamy-white, sweet-scented. Calyx of 5 strongly imbricate free sepals, these 2·5–3 mm. long, oblong, rounded or emarginate at apex, characteristically glistening reddish-brown at least when dry, margin ciliate, dorsal surface glabrous or the midrib finely long-hispid; calyx $\frac{1}{2}$–$\frac{2}{3}$ the length of the corolla tube. Corolla tube 3·5–4·5 mm. long, clavate, its wall thickened at the level of the anthers, its mouth constricted; outer surface finely puberulent, giving a mealy appearance; corolla lobes narrowly elliptic, ± equalling tube in length, dorsally mealy-puberulent on the overlapping surface. Stamens inserted in the upper half of the tube with the anthers reaching to the mouth; anthers c. 1 mm. long. Ovary c. 1 mm. long, cylindrical, densely pubescent at apex; style, clavuncle and stigma 1·5–2 mm. long. Fruit up to 5 cm. in diameter, globose, white or purplish or greenish with brown spots, or brown or purplish-blue, edible. Seeds 14–26 mm. long.

Zambia. N: Mporokoso Distr., Kundabwika Falls, Kalungwishi R., 945 m., fl. 19.ix.1957, *Whellan* 1413 (K; PRE; SRGH). W: Solwezi Distr., Chifubwa Gorge c. 5 km. S. of Solwezi, c. 1310 m., fl. 19.xii.1969, *Simon & Williamson* 1856 (K; LISC; PRE; SRGH). C: between Rufunsa and Lusaka, 1180 m., fr. 26.iii.1955, *E.M. & W.* 1208 (BM; LISC; SRGH). S: Namwala Distr., 8·3 km. N. of Ngoma, Kafue National Park, fl. 17.i.1963, *Mitchell* 17/39 (FHO; K; SRGH). **Malawi.** N: Nkhata Bay, Chikale Beach, 475 m., st. 13.viii.1972, *Pawek* 5648 (K; SRGH). C: Nkhota Kota Distr., Chia area, 480 m., fr. 5.ix.1946, *Brass* 17535 (K; SRGH). **Mozambique.** N: Cabo Delgado, Macondes, c. 3 km. from Mueda to Nantulo, c. 800 m., fr. 15.iv.1964, *Torre & Paiva* 12049 (LISC; LMA).

Also in Zaire, Tanzania, Burundi and Angola. In open bush or woodland, on rocky ground.

Pichon (loc. cit.) treats this taxon as var. *parvifolia* of the aggregate species *L. parvifolia* which also includes var. *johnsonii* (A. Chev.) Pichon (Nigeria to Zaire and Angola) and var. *thollonii* (Dewèvre) Pichon (Congo Rep., Zaire and Angola).

5. **Landolphia owariensis** P. Beauv., Fl. Oware Bénin **1**: 55, t. 34 (1806). — Pichon in Mém. Inst. Fr. Afr. Noire **35**: 109, map 7 (1953). — H. Huber in F.W.T.A., ed. 2, **2**: 55, fig. 213 (1963). — Fanshawe, Check List Woody Pl. Zambia: 25 (1973). Type from Nigeria.

Vahea owariensis (P. Beauv.) F. Muell. in Wittst., Org. Const. Pl.: 258 (1878). Type as above.
Pacouria owariensis (P. Beauv.) Hiern in Cat. Afr. Pl. Welw. 1: 661 (1898). Type as above.
Landolphia gentilii De Wild., Obs. Apocyn. Latex: 20 (1901). Type from Zaire.
Landolphia owariensis var. *tomentella* Stapf in F.T.A. 4, 1: 51 (1902). Syntypes from the Sudan.
Landolphia humilis K. Schum. ex Stapf, tom. cit.: 53 (1902). Type from Zaire.
Landolphia pierrei Hua in C.R. Acad. Sci. Paris 135: 868 (1902). Lectotype from Gabon.
Landolphia stolzii Busse in Engl., Bot. Jahrb. 32: 168 (1902). Type from Tanzania.
Landolphia droogmansiana Wildem. in Wildem. & Gentil, Lianes Caoutch. Congo: 59 (1904). Type from Zaire.
Landolphia tomentella (Stapf) A. Chev. in Bull. Soc.Bot. Fr. 53: 21 (1906). Type as for *Landolphia owariensis* var. *tomentella*.
Landolphia turbinata Stapf ex A. Chev., tom. cit.: 32 (1906). Type from Uganda.
Landolphia glaberrima A. Chev., op. cit. 55, Mém. 8: 45 (1908). Syntypes from Central African Republic.
Landolphia mayumbensis Good in Journ. Bot., Lond. 67, Suppl. Gamop.: 83 (1929). Type from Angola.
Ancylobothrys sp. 1. — F. White, F.F.N.R.: 346 (1962).

Liane climbing to 12 m. or more; young twigs ferrugineous-pubescent, indumentum later becoming greyish, rather persistent; bark on young branches dark reddish brown with tiny circular white lenticels. Leaves thinly coriaceous; petiole c. 5 mm. long, pubescent all round; lamina 7·5–11·6 × 3–4·4 cm., oblong, apex shortly acuminate, acumen rounded, base rounded. Upper surface glossy, brownish-green when dry, glabrous, midrib impressed, other nerves raised or level; lower surface mat, pale brownish when dry, lamina glabrous or sparsely pubescent near the petiole, midrib subglabrous to hispid, all nerves raised. Tendrils uncommon. Inflorescences terminal many-flowered capitate or lax cymose panicles on peduncles 0–1·8 cm. long; axes densely ferrugineous-pubescent. Flowers white or cream-coloured, sweet-scented. Calyx 2–2·5 mm. long, lobes broadly elliptic, densely appressed ferrugineous-pubescent, the indumentum characteristically parted along the median line. Corolla tube 6–6·5 mm. long, swollen in the upper half but the wall not thickened, constricted at the mouth, puberulent on outer surface and pubescent internally above the stamens; corolla lobes 2–3·5 × 1–1·5 mm., elliptic, subglabrous on outer surface and glabrous within. Stamens inserted in the upper part of the tube so that the anthers reach to 1 mm. of the mouth; anthers subsessile, c. 0·75 mm. long. Ovary c. 0·5 mm. high, compressed-globose, pubescent; style, clavuncle and stigma 3–3·5 mm. long. Fruit up to 15 cm. in diameter, globose, yellow or green turning brown, inedible. Seeds 10–18 mm. long.

Zambia. N: Mbala (Abercorn) Distr., L. Chila, 1550 m., fl. 6.xi.1958, *Robson & Fanshawe* in *Robson* 501 (BM; BR; K; LISC; SRGH). W: Mwinilunga Distr., Zambezi Rapids, fl. 18.v.1969, *Mutimushi* 3379 (K). **Malawi.** N: Nkhata Bay, st. 14.vi.1954, *Jackson* 1344 (BR; FHO; K).
Widely distributed in W. tropical and Central Africa. In riparian woodland and mushitu.

As Pichon (1953) has emphasised, this is the most widespread and variable species of *Landolphia*. In two areas of Central Africa, separated by equatorial forest (Pichon, op. cit., map 7), it is found in savanna habitats as a fire-resistant suffrutex less than 1 m. high. In dense forest, on the other hand, its stems can reach 100 m. in length.

6. **Landolphia eminiana** Hallier f. in Jahrb. Hamb. Wiss. Anst. 17, Beih. 3: 88 (1900). — Pichon in Bull. Jard. Bot. Brux. 22: 107 (1952); in Mém. Inst. Fr. Afr. Noire 35: 137, t. 3 fig. 7–8, map 8 (1953). — Fanshawe, Check List Woody Pl. Zambia: 25 (1973). Type from Tanzania.

Scandent shrub climbing to 12 m. or more. Young stems glabrous. Leaves thinly coriaceous; petiole 4–6 mm. long, glabrous below, reddish-pilose above; lamina 3·5–10·5 × 1·6–4 cm., oblong-elliptic, acute to attenuate-acuminate at apex (acumen absent or short to long, with much variation even on the same twig), base obtuse. Upper leaf surface characteristically mat grey when dry, glabrous, midrib impressed, lateral nerves and reticulum raised; lower surface pale brownish when dry, glabrous, all nerves raised. Tendrils uncommon. Inflorescences dense capitate many-flowered terminal cymose panicles; peduncle 0–9 mm. long; axes ferrugineous-pilose. Flowers sweet-scented, white with external indumentum pale buff. Calyx 3–5 mm. long, lobes elliptic to broadly spathulate with patent tips, free to base, strongly imbricate, dorsally densely ferrugineous appressed-pilose especially along the median line. Corolla tube 17–22 mm. long, slender, very slightly inflated in the middle, external surface densely pubescent, inner surface pubescent just below the stamens; corolla lobes 10–15 × 3·5–8 mm., elliptic, glabrous on upper surface, densely pubescent below on exposed half. Stamens inserted at the middle of the corolla tube; anthers subsessile, c. 2 mm. long. Ovary c. 1 mm. long, gynoecium 7–8 mm. long; ovary and lower ⅔ of the style densely long-pilose. Fruit 3·5–4·5 cm. in diameter, subglobose or oblong-pyriform. Seeds 9–11·5 mm. long.

Zambia. N: Mbala (Abercorn) Distr., Vomo Gap, Fwambo side, 1500 m., fl. 27.ix.1960, *Richards* 13274 (K; SRGH). W: Kitwe Distr., S. Mutundu Botanical Reserve near Mufulira, fl. 13.x.1970, *Fanshawe* 10932 (K).
Also known from Zaire, Rwanda and Tanzania. In dry evergreen forest patches, dry river beds and on margin of riverine vegetation.

7. **Landolphia rufescens** (De Wild.) Pichon in Mém. Inst. Fr. Afr. Noire **35**: 161, t. 6 fig. 1–2, map 10 (1953). — F. White, F.F.N.R.: 350 (1962). — Fanshawe, Check List Woody Pl. Zambia: 25 (1973). Type from Zaire.
    Carpodinus rufescens De Wild., Not. Pl. Ut. Congo **2**: 241 (1908). Type as above.

Liane ascending to 24 m. or more. Young stems dark brown, hispid-pubescent, sometimes sparsely so. Leaves moderately coriaceous, often drying blackish-green; petiole 3–10 mm. long, with indumentum like the stem; lamina 5·5–15·5 × 2·9–7·4 cm., elliptic to ovate- or obovate-elliptic, the apex acuminate to a long narrow round-tipped or acute acumen or rarely rounded and only faintly apiculate, the base slightly cordate or less often cuneate. Upper leaf surface entirely glabrous or sparsely pilose on midrib; midrib broad, channelled, lateral nerves and tertiary reticulation impressed; lower surface with lamina glabrous and midrib and lateral nerves brownish hispid-pubescent, sometimes only sparsely; all veins prominent, the loops joining the lateral nerves conspicuous, well spaced in from the margin. Tendrils present. Inflorescences axillary few- to several-flowered cymes, sessile or subsessile, with short branches. Flowers yellowish, sweet-scented, subtended by two sepal-like bracteoles. Calyx c. 2 mm. long, sepals ovate, appressed-pubescent, free, patent. Corolla tube 3–6 mm. long, puberulous on outer surface, glabrous within, inflated in upper half; corolla lobes 2·5–4 mm. long, linear, puberulous on outer surface, glabrous above. Stamens inserted in upper half of corolla tube; anthers c. 1 mm. long. Gynoecium narrowly conical; ovary glabrous at base, appressed-pubescent above, passing smoothly into the glabrous style. Fruit globose, up to 4 cm. in diameter, smooth, pale green with longitudinal markings round the apex. Seeds c. 8 mm. long.

Zambia. W: Mwinilunga Distr., 97 km. from Mwinilunga to Solwezi, fr. 18.ix.1952, *Angus* 479 (BM; BR; FHO; K; PRE).
Also from Zaire and Angola. In mushitu vegetation.

Reported by Fanshawe (loc. cit.) from the Northern Province of Zambia.

8. **Landolphia cuneifolia** Pichon in Mém. Inst. Fr. Afr. Noire **35**: 163, t. 6 fig. 3, map 10 (1953). TAB. **92**. Type from Zaire.
    Landolphia gossweileri (Stapf) Pichon, tom. cit.: 186, map 12 (1953) pro parte quoad *Trapnell* 1587 — sensu F. White, F.F.N.R.: 349 (1962); in Gard. Bull., Singapore **29**: 68 (1976). — sensu Fanshawe, Check List Woody Pl. Zambia: 25 (1973).

Suffrutex or climbing shrub up to 1 m. (or more?) high. Young stems dark brown, hispid or pubescent. Leaves coriaceous, characteristically yellowish-green; petiole 2–6 mm. long, pubescent; lamina 4·5–8·5 × 0·8–4 cm., elliptic, obovate-elliptic or oblong-elliptic, the apex cuspidate to a well-marked short to moderately long acumen with apex rounded to acute, the base cuneate to obtuse. Upper leaf surface glossy, glabrous except for the hispidulous midrib, midrib deeply impressed, lateral nerves faintly so, tertiary reticulation obscure; lower leaf surface mat, lamina with very sparse scattered hairs, midrib hispidulous; midrib and lateral nerves prominent, reticulation inconspicuous; lamina with irregularly scattered glandular dots, or these sometimes absent. Lateral nerves 7–9-paired. Tendrils present but uncommon. Infloresences all axillary, flowers solitary or in few-flowered sessile fascicles, subtended by sepal-like bracts; flowers white, cream or lemon-yellow. Calyx c. 2

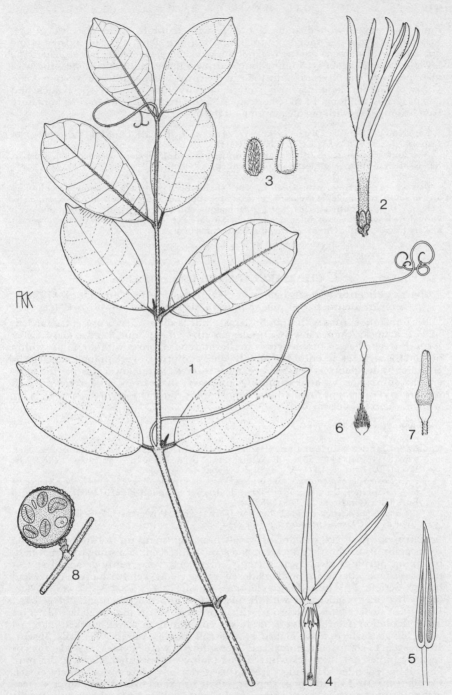

Tab. 92. LANDOLPHIA CUNEIFOLIA. 1, habit (× ⅔), from *Drummond & Rutherford-Smith* 7025;
2, flower bud with corolla lobes partly opened (anomalous 6-merous flower) (× ⅚); 3, calyx
lobe, dorsal and ventral views (× 4); 4, part of corolla opened out (× ⅘); 5, stamen (× 20); 6,
gynoecium (× 4); 7, detail of apex of gynoecium (× 20), 2–7 from *Hooper & Townsend* 287;
8, fruit in cross-section, showing seeds (× ⅔), from *Homblé* 740.

mm. long, sepals united at base, oblong with truncate apiculate apex, ciliate, hispidulous along the midrib. Corolla tube c. 8 mm. long, slender, inflated at the apex at level of stamens but the wall not thickened here, externally puberulent in the lower half; corolla lobes c. 11 mm. long, narrowly linear, minutely ciliate, otherwise glabrous. Stamens inserted in the top $\frac{1}{4}$ of the corolla tube, anthers c. 1·5 mm. long. Ovary c. 1·5 mm. long, narrowly conical, apically pubescent; style, clavuncle and stigma c. 5·5 mm. long. Fruit spherical, c. 2·5 cm. in diameter when ripe; immature fruit bright green with purple markings at the base.

Zambia. W: Solwezi Distr., 19 km. W. of Solwezi, fl. buds 18.iii.1961, *Drummond & Rutherford-Smith* 7025 (K; LISC; PRE; SRGH).
Also known from Zaire (Shaba). In *Brachystegia* and *Marquesia* woodland on Kalahari Sands, often forming an almost pure understorey.

Landolphia gossweileri, with which *L. cuneifolia* has been confused, is confined to a limited area of southern Angola. It is closely related to the common and widespread *L. lanceolata* (K. Schum.) Pichon, both species being rhizomatous suffrutices with no tendrils and with verticillate leaves. The two differ chiefly in their inflorescences, *L. gossweileri* having flowers in axillary fascicles and *L. lanceolata* having lax terminal cymes.

4. CHAMAECLITANDRA (Stapf) Pichon

Chamaeclitandra (Stapf) Pichon in Mém. Inst. Fr. Afr. Noire **35**: 202 (1953).
    Clitandra Benth. Sect. *Chamaeclitandra* Stapf in F.T.A. **4**, 1: 61 (1902).

Rhizomatous shrubs with straight stems; tendrils absent. Latex present, abundant in the rootstocks. Spines absent. Stipules absent. Cymes terminal and axillary. Calyx lobes ± free to base. Corolla hypocrateriform; tube glabrous without, pilose within around the stamens and near the mouth; lobes contorted, overlapping to the left, ± equalling tube, glabrous or ciliate. Stamens inserted in the lower third of the tube, anthers subsessile, without carina. Ovary hairy; ovules few, in 2 series in each loculus; style, clavuncle and stigma very short. Fruit a few-seeded glabrous berry without sclerified pericarp.

A monotypic African genus.

Chamaeclitandra henriquesiana (Hallier f.) Pichon in Mém. Inst. Fr. Afr. Noire **35**: 203, t. 9 fig. 8–11, map 14 (1953). — F. White, F.F.N.R.: 347 (1962). — Fanshawe, Check List Woody Pl. Zambia: 10 (1973). TAB. **93**. Type from Angola.
    Clitandra henriquesiana K. Schum. ex Warb. in Tropenpfl. **1**: 134 (1897), *nom. nud. cum ic.* — Stapf in F.T.A. **4**, 1: 62 (1902); in Journ. Bot., London **46**: 210 (1908). — Pichon in Bull. Jard. Bot. Brux. **22**: 108 (1952).
    Landolphia henriquesiana Hallier f. in Jahrb. Hamb. Wiss. Anst. **17**, Beih. 3: 97 (1900). Type as for *Chamaeclitandra henriquesiana*.

Suffrutex with erect, straight, sparingly branched stems up to 1 m. high. Young stems glabrous or shortly pubescent and soon glabrescent, chestnut-brown at first, becoming purple-brown. Leaves thinly coriaceous, very neatly arranged and of uniform shape, characteristically deflexed on the stems, yellowish green above and below; petiole 3–4 mm. long, glabrous; lamina 3·4–5·5 × 1·3–2·3 cm., ovate, narrowly ovate or lanceolate with distinct round-tipped acumen and obtuse base. Upper leaf surface glossy, glabrous, with impressed midrib and raised lateral nerves and reticulation; lateral nerves closely spaced, joining in a well-marked marginal vein; lower surface mat, glabrous, with all nerves raised. Tendrils absent. Inflorescences compact, terminal and lateral, few-flowered cymes; axes glabrous or sparsely pubescent. Flowers white tinged pink, sweet-scented. Calyx c. 1·5 mm. long, lobes rounded-triangular, free to base, dorsally glabrous, minutely ciliate. Corolla tube 10–13 mm. long, slender, inflated in the lowest third, outer surface glabrous, inner surface shortly pilose in the lower half and near the mouth; corolla lobes ± equalling tube, narrowly elliptic, glabrous or inner margins sometimes ciliate. Stamens inserted 2–3·5 mm. from the base of the corolla tube, anthers 1 mm. long. Ovary subglobose, 1 mm. long, minutely pubescent at apex. Style, clavuncle and stigma 1 mm. long. Fruit up to 6·5 × 3·5 cm., pyriform or obovoid, green, yellow or orange with scattered brownish lenticels, the surface somewhat rugose, edible with refreshing taste. Seeds 3–4 per fruit, c. 3 cm. long.

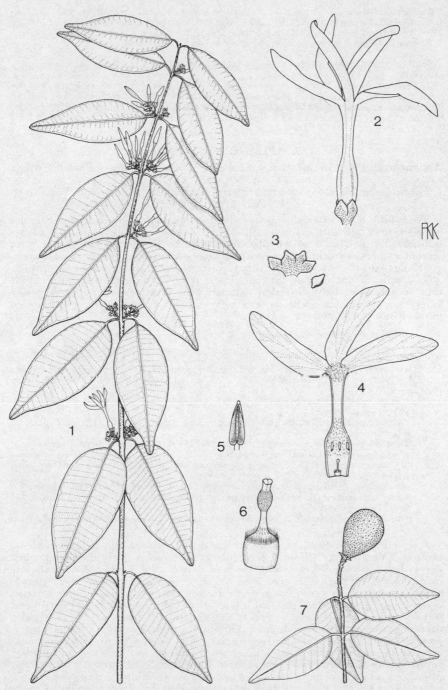

Tab. 93. CHAMAECLITANDRA HENRIQUESIANA. 1, habit (× ⅔); 2, flower (× ⅖); 3, calyx (× ⅖); 4, part of flower, opened out (× ⅖); 5, stamen (× 12); 6, gynoecium (× 12); 7, apex of stem with young fruit (× ⅔). 1–6 from *Richards* 16896; 7 from *Brenan & Keay* 7704.

Zambia. B: Mongu Distr., c. 70 km. from Mongu to Kaoma (Mankoya), fl. 19.xi.1959, *Drummond & Cookson* 6607 (K; LISC; PRE; SRGH). W: Mwinilunga Distr., Kalene Hill, fl. 25.ix.1952, *Angus* 547a (BR; FHO; K). S: Namwala, fl. 20.x.1959, *Fanshawe* 5256 (FHO; K). Known also from Angola and Zaire. In woodland on Kalahari Sands.

Stapf (loc. cit., 1908) gives fruit dimensions of up to 9·5 × 5·5 cm., and cites Gossweiler's observation of the flower colour as blue.

The earliest name for this species is *Clitandra henriquesiana* K. Schum. ex Warb., but the latter cannot be regarded as validly published because there is no description and the illustration is unaccompanied by any kind of analysis (cf. ICBN, Art. 32 and 44, 1978). In fact, the caption refers the name *Clitandra henriquesiana* to the figure labelled D, in error for E.

5. APHANOSTYLIS Pierre

Aphanostylis Pierre in Bull. Soc. Linn. Paris, N.S. **1**: 89 (1898). — Pichon in Mém. Inst. Fr. Afr. Noire **35**: 235 (1953).

Clitandra Benth. Sect. *Aphanostylis* (Pierre) Hallier f. in Jahrb. Hamb. Wiss. Anst. **17**, Beih. 3: 122 (1900).

Tendrillous lianes. Latex present, sometimes abundant. Spines absent. Stipules absent. Cymes both terminal and axillary (in F.Z. area) or only axillary. Calyx lobes imbricate, free to base. Corolla hypocrateriform; tube ± cylindrical, dilated or not at level of anthers, sometimes hairy within; lobes contorted, overlapping to the left, equalling to greatly exceeding the tube, sometimes hairy above at the base, with or without cilia on inner margin. Stamens inserted at various levels within the corolla tube; anthers often slightly exserted, without carina. Ovary glabrous or hairy; ovules several–∞; style short, glabrous; clavuncle annular. Fruit a berry with sclerified pericarp (not in F.Z. area) or without.

An African genus comprising 3 species.

Aphanostylis mannii (Stapf) Pierre in Bull. Soc. Linn. Paris, N.S. **1**: 89 (1898). — Pichon in Mém. Inst. Fr. Afr. Noire **35**: 241, t. 13 fig. 3–5, map 16 (1953). — F. White, F.F.N.R.: 347 (1962). — H. Huber in F.W.T.A., ed. 2, **2**, 59 (1963). TAB. **94**. Type from Sierra Leone.
    Clitandra mannii Stapf in Kew Bull. **1894**: 20 (1894). — Irvine, Woody Pl. Ghana: 620 (1961). Type as above.
    Carpodinus exserens K. Schum. in Engl., Bot. Jahrb. **23**: 219 (1896). Type from Cameroon.
    Carpodinus laxiflora K. Schum., tom. cit.: 220 (1896). Type from Sierra Leone.
    Carpodinus mannii (Stapf) Stapf ex K. Schum., loc. cit. Type as for *Aphanostylis mannii*.
    Aphanostylis exserens (K. Schum.) Pierre, tom. cit.: 90 (1898). Type as for *Carpodinus exserens*.
    Aphanostylis laxiflora (K. Schum.) Pierre, loc. cit. Type as for *Carpodinus laxiflora*.
    Clitandra laxiflora (K. Schum.) Hallier f. in Jahrb. Hamb. Wiss. Anst. **17**, Beih. 3: 124 (1900). Type as above.
    Clitandra gentilii De Wild., Not. Apocynac. Latic. Fl. Congo **1**: 24 (1903). Type from Zaire.

Liane climbing to 40 m. or more, or scandent shrub sometimes forming dense thickets. Young stems glabrous. Leaves thinly coriaceous, scarcely discolorous when dry, glabrous; petiole 5–8 mm. long; lamina 5·2–10·6 × 2·4–5 cm., oblong or slightly ovate-oblong, contracted at apex into a distinct, narrow, parallel-sided, round-tipped acumen up to 9 mm. long; upper surface mat, midrib raised, other nerves conspicuous; lower surface mat, midrib level or slightly raised, lateral nerves and tertiary reticulation slightly in relief. Venation with lateral nerves ± straight, diverging from midrib at a wide angle, fairly well spaced, joined by shallow loops ± parallel to the margin. Tendrils terminal, long, woody, with recurved hooks representing sterile inflorescence branches. Inflorescences terminal and axillary several–many-flowered cymes up to 18 mm. long with slender, glabrous or sparsely pilose axes and conspicuous sepal-like bracts. Flowers white, tinged pink in bud, sweet-scented. Calyx c. 1 mm. long, lobes ovate, dorsally glabrous, ciliate. Corolla tube 2–2·5 mm. long, glabrous externally, sometimes pubescent within; corolla lobes ± equalling tube, narrowly elliptic, glabrous or pubescent on upper surface, with or without marginal cilia. Stamens inserted in upper half of corolla tube, anthers c. 1 mm. long, sometimes reaching mouth of tube. Ovary c. 1 mm. long, conical; style very short; gynoecium c. 1·5 long. Fruit ovoid and green when young, swelling to

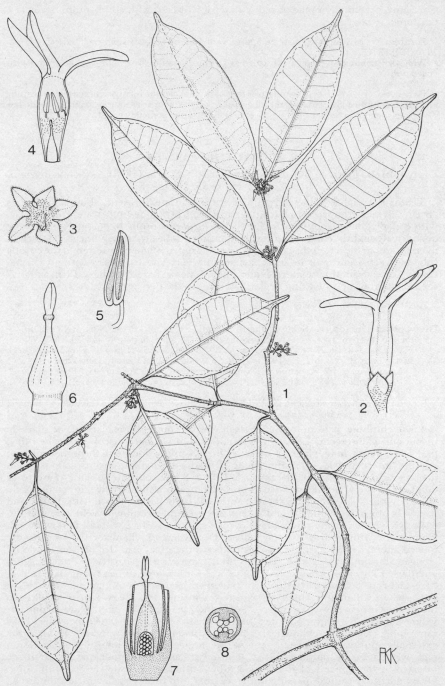

Tab. 94. APHANOSTYLIS MANNII. 1, habit (× ⅔); 2, flower (× 6); 3, calyx (× 6); 4, part of corolla opened out (× 6); 5, stamen (× 20); 6, gynoecium (× 20); 7, longitudinal section through lower part of flower (× 14); 8, diagrammatic transection through ovary. All from *White* 3370.

4 × 2 cm., becoming orange or red with dark lenticels, with edible pulp. Seeds c. 1 cm. long.

Zambia. W: Mwinilunga Distr., 6·5 km. N. of Kalene Hill Mission, fl. 25.ix.1952, *White* 3370 (BR; FHO; K).
Widely distributed in tropical Africa from Guiné-Bissau to Zaire and Angola. In mushitu vegetation.

Pichon, op. cit.: 244 (1953), emphasises that in this species the relative lengths of corolla tube and petals are unusually variable; the lobes can be up to six times as long as the tube. In the few flowering specimens seen from the F.Z. area they are approximately equal.

6. DICTYOPHLEBA Pierre

Dictyophleba Pierre in Bull. Soc. Linn. Paris, N.S. **1**: 92 (1898). — Pichon in Mém. Inst. Fr. Afr. Noire **35**: 250 (1953).

Tendrillous lianes with interpetiolar stipules. Latex present, abundant. Spines absent. Cymes grouped into terminal panicles. Calyx lobes imbricate, free to base, with conspicuous glandular auricles (in *D. lucida*). Corolla hypocrateriform; tube cylindrical, slightly enlarged at level of anthers; lobes narrowly linear, contorted, overlapping to the left, long-ciliate on inner margins. Stamens inserted in the corolla tube between ¼ and ¾ of its length; filaments very short; anthers without carina. Ovary glabrous, multiovulate; clavuncle positioned at base of anthers. Fruit a many-seeded glabrous berry lacking a sclerified layer in the pericarp.

An African genus of 4 species.

Dictyophleba lucida (K. Schum.) Pierre in Bull. Soc. Linn. Paris, N.S. **1**: 93 (1898). — Pichon in Mém. Inst. Fr. Afr. Noire **35**: 262, t. 15 fig. 4–6, map 18 (1953). — F. White, F.F.N.R.: 348 (1962). — H. Huber in F.W.T.A., ed. 2, **2**: 59 (1963). — Fanshawe, Check List Woody Pl. Zambia: 15 (1973). TAB. **95**. Lectotype (of Pichon) from Zaire.
    Landolphia lucida K. Schum. in Notizbl. Bot. Gart. Berl. **1**: 24 (1895). Type as above.
    Pacouria lucida (K. Schum.) Kuntze in Deutsche Bot. Monatsschr. **21**: 173 (1903). Type as above.
    Landolphia dubreucquiana De Wild. in De Wild. & Gentil, Lianes Caoutch. Congo: 92 (1904). Syntypes from Zaire.

Liane climbing to 15 m. or more; slash white. Young stems glabrous or sparsely hispid and glabrescent. Leaves membranous, not strongly discolorous when dry; petiole 3–10 mm. long, glabrous or with a hispid-pilose line along the upper surface; stipules interpetiolar, 1–1·5 mm. long, broadly triangular, inconspicuous, soon disintegrating or caducous; lamina 5·5–14 × 3–8 cm., obovate, oblong-obovate or elliptic, the apex rounded or with a short obscure to well-marked acumen or rarely slightly emarginate, the base shortly auricled. Upper leaf surface with deeply and narrowly impressed midrib and lateral nerves and prominent vein reticulation, entirely glabrous or the midrib hispid-pilose; lower surface with all nerves raised, glabrous or with midrib sparsely pilose. Leaf venation characterised by strongly looped lateral nerves forming a continuous submarginal vein. Inflorescence a loose elongate terminal panicle of few-flowered cymes with sensitive peduncles; axes glabrous. Flowers sweet-scented. Calyx 2–2·5 mm. long, lobes triangular, free to base, glabrous but minutely ciliate, provided with conspicuous glandular auricles at the base of each margin which overlaps to the inside (thus the two inner sepals each has 2 auricles, the two outer have none and the fifth has one). Corolla pink or yellowish in bud, becoming white at anthesis with pale to dark pink tube; tube 11–19 mm. long, narrowly cylindrical, slightly dilated in the upper ½–⅓, glabrous externally, pubescent internally below the stamens; corolla lobes 9–14 mm. long, narrowly linear, strongly contorted, long-ciliate along the inner margin. Stamens inserted in the upper ⅓ of the corolla tube; anthers subsessile, 1·75–2 mm. long. Ovary narrowly conical, passing imperceptibly into the style; clavuncle cylindrical; gynoecium 8–10 mm. long. Fruit up to 3 × 4·5 cm., ± spherical, green to yellow or red with yellow edible pulp. Seeds up to c. 11 mm. long.

Zambia. N: Kawambwa Distr., around Kafulwe Mission, Lake Mweru, fl. 6.xi.1962, *White* 3628 (BR; FHO; K). W: Mwinilunga, fl. buds 11.iii.1962, *Holmes* 1494 (K). W/C: Ndola Distr., Miengwe Forest Reserve, st. 24.iv.1942, *R. G. Miller* 344 (FHO). **Zimbabwe.** E:

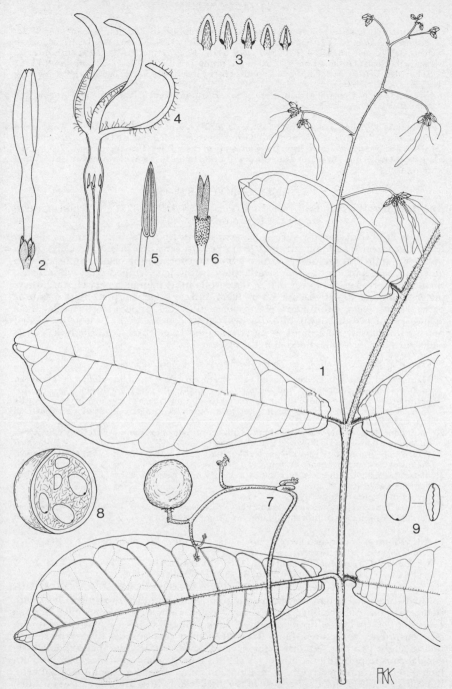

Tab. 95. DICTYOPHLEBA LUCIDA. 1, habit (×⅔), from *Pawek* 6418 and *Garcia* 296; 2, flower bud
(×2); 3, calyx showing dark auricles on the covered margins (×4); 4, part of flower opened
out (×2); 5, stamen (×12); 6, apex of gynoecium (×12), 2–6 from *Barbosa* 1252; 7,
infructescence with young fruit (×⅘); 8, fruit, semidiagrammatic reconstruction showing
seeds embedded in fibrous matrix (×⅔); 9, seed, dorsal and lateral views (×1), 7–9 from
Torre & Correia 1437.

Chimanimani (Melsetter) Distr., Haroni/Makurupini Forest, 400 m., fl. xii.1964, *Wild, Goldsmith & Müller* 6617 (BR; FHO; K; SRGH). **Malawi.** N: Nkhata Bay Distr., Mzuzu to Nkhata Bay, c. 610 m., fl. 4.ii.1973, *Pawek* 6418 (K; MAL; SRGH). **Mozambique.** Z: Maganja da Costa, Gobene Forest, c. 20 m., fr. immat. 14.ii.1966, *Torre & Correia* 14611 (C; LISC; LMA; WAG). MS: Chimoio, Bandula Distr., Jagersberg Farm, fl. 28.i.1949, *Chase* 999 (BM; K; LISC; SRGH).

Nigeria, Gabon, Central African Rep., Zaire, Burundi, Kenya, Tanzania and Angola (the extreme NW). In hygrophilous forest.

At the time of Pichon's monograph (1953), *D. lucida* was unknown from Kenya, Zambia and Malawi, and the populations which had been discovered in Zimbabwe (E) and Mozambique (MS) seemed widely separated from the main body of the species (Pichon, tom. cit.: 265 & 371). More recent collections have filled out the distribution range so that it conforms with a common pattern.

7. ANCYLOBOTRYS Pierre

Ancylobotrys Pierre in Bull. Soc. Linn. Paris, N.S. **1**: 91 (1898). — Pichon in Mém. Inst. Fr. Afr. Noire **35**: 272 (1953).

Tendrillous lianes or low shrubs. Latex present. Spines absent. Stipules absent. Cymes grouped into long terminal panicles with sensitive branches. Calyx lobes imbricate, united at the base. Corolla hypocrateriform; tube usually pubescent on outer surface, lobes contorted, overlapping to the left, fringed with white hairs. Stamens inserted in the lower half of the corolla tube; filaments very short, anthers introrse, dorsifixed, with carina. Ovary hairy, uni- or bilocular, placentation axile or parietal with very prominent placentae, ovules ∞. Style glabrous; clavuncle cylindrical or ovoid; stigma bifid but with arms connivent. Fruit a many-seeded berry with densely velutinous surface; sclerified layer absent; pulp edible. Seeds endospermous, firmly embedded in the placental pulp.

A genus of 10 species, all occurring in Africa and one also in Madagascar.

1. Main lateral leaf veins closely spaced (1·5–4 mm. apart), ± straight, almost at 90° to the midrib, connected in a neatly and symmetrically scalloped submarginal vein   -   -   2
 – Main lateral leaf veins more distantly spaced (5–20 mm. apart or more), curving towards apex, looped to join with the next, thus forming an asymmetrically scalloped vein well spaced from the margin   -   -   -   -   -   -   -   -   -   -   -   3
2. High liane; leaf lamina 4–8 × 1·7–3 cm., with acuminate apex; main inflorescence branches ending in a dense many-flowered cymose cluster   -   -   -   -   -   2. *tayloris*
 – Low much-branched twiggy shrub; leaf lamina 1·4–3·8 × 0·8–1·6 cm., with rounded apex; main inflorescence branches ending in a 1–3 flowered cyme   -   -   -   - 1. *capensis*
3. Stems persistently densely ferrugineous-velutinous; petioles 9–12 (15) mm. long; leaf lamina usually acute at apex -   -   -   -   -   -   -   -   - 4. *amoena*
 – Stems and leaves not as above   -   -   -   -   -   -   -   -   -   4
4. Petioles 3–8 mm. long; leaf lamina rounded to obtuse at apex   -   -   - 3. *petersiana*
 – Petioles 17 mm. long or more; leaf lamina acuminate   -   -   -   - 5. *pyriformis*

1. **Ancylobotrys capensis** (Oliv.) Pichon in Mém. Inst. Fr. Afr. Noire **35**: 297, t. 19 fig. 4–5, map 20 (1953). Lectotype from S. Africa (Transvaal).
    Landolphia capensis Oliv. in Hook., Ic. Pl. **13**: t. 1228 (1877). — Stapf in F.C. **4**, 1: 495 (1907). — Codd in Fl. Southern Afr. **26**: 261 (1963). Type as above.
    Pacouria capensis (Oliv.) S. Moore in Journ. Bot., Lond. **41**: 403 (1903). Type as above.

Much-branched twiggy shrub, creeping along the ground or growing to 3 m., with stems farinose-puberulous. Leaves coriaceous; petiole 3–6 mm. long; lamina 1·4–3·8 × 0·8–1·6 cm., elliptic to oblong, apex rounded to obtuse, without acumen, base cuspidate, sometimes fairly tapered. Upper leaf surface mat, puberulous, midrib slightly impressed, other nerves level; lower surface mat, paler than above, subglabrous to puberulous, all nerves slightly raised. Venation characteristic, with lateral nerves running straight to margin and joined in a looped submarginal vein, but veins relatively obscure because of the thick leaf texture. Inflorescences terminal panicles in which the few branches end in 1–3-flowered cymes; flowers also present, solitary and sessile, in axils of leaf-pair below the peduncle. Flowers white, sweet-scented. Calyx 4–5 mm. long, lobes narrowly linear, rounded at apex, united at base, dorsally densely ferrugineous appressed-pubescent. Corolla tube 7–10 mm. long, slender, slightly inflated just below the middle, outer surface densely ferrugineous-pubescent above the level of the calyx, the indumentum continuing onto the lobes, inner surface of tube pubescent in the upper half. Corolla lobes 7–17 mm. long, 1–1·5 or more times as long as tube, lobes narrowly obovate, glabrous on inner surface, $\pm$ pubescent externally, shortly ciliate. Stamens inserted c. 2·5 mm. from the base of the tube; anthers subsessile, c. 1·5 mm. long. Ovary c. 0·75 mm. long, globose, pubescent. Style, clavuncle and stigma c. 2 mm. long. Fruit subglobose, up to 5 cm. in diameter. Seeds 10–12 mm. long.

Botswana SE: Nuane Dam, NW of Lobatsi, fl. 26.ix.1964, *Leach, Bayliss & Lamont* 12455A (K; SRGH).
Also found in S. Africa (Natal and Transvaal). On rocky hillsides.

2. **Ancylobotrys tayloris** (Stapf) Pichon in Mém. Inst. Fr. Afr. Noire **35**: 284, t. 18 fig. 1–3, map 20 (1953). TAB. **96** fig. A. Type from Kenya.
  Landolphia tayloris Stapf in F.T.A. **4**, 1: 45 (1902). Type as above.
  Landolphia pachyphylla Stapf, loc. cit. Type: Malawi, "Nyasaland", 1895, *Buchanan* 140 (BM, holotype).

Liane. Young stems puberulent, tardily glabrescent. Leaves coriaceous, drying grey-green to dark brown, paler below; petiole 2–9 mm. long, with indumentum like the stem; lamina 4–8 × 1·7–3 cm., oblong-elliptic, apex with short acute acumen often deflexed (so that the leaf cannot be pressed flat without folding), base rounded-cuspidate. Upper leaf surface glossy, puberulent, glabrescent, with midrib and lateral nerves impressed, reticulation obscure; lower surface mat, completely glabrous or with appressed-pubescent midrib and sparsely pubescent lamina, all nerves raised, the laterals characteristically closely spaced, running straight out to the margin and joining in a well-marked looped submarginal vein, the intervening areas filled with a neat conspicuous tertiary reticulation. Inflorescences elongate panicles, each lateral branch ending in a dense cymose flower-cluster, axes densely rufous-puberulous. Flower buds pink, flowers white, sweet-scented. Calyx 2–3 mm. long, the lobes narrowly triangular, united at the base, dorsally densely ferrugineous appressed-pubescent. Corolla tube 9–10·5 mm. long, slender, slightly inflated in the lower half, outer surface pubescent, inner surface glabrous except for sparse pubescence above the anthers; corolla lobes 1–1·5 times as long as tube, narrowly elliptic, with acute apex; both surfaces glabrous, margins ciliate with crispate white hairs. Stamens inserted 1·5–2·5 mm. from base of corolla tube; anthers subsessile, 0·75–1·25 mm. long. Ovary c. 0·75 mm. long, globose, pubescent. Style, clavuncle and stigma 1–1·5 mm. long. Fruit c. 4 cm. in diameter, yellow, with edible pulp.

Malawi. S: Mulanje Mt. Forest, beside Lukulezi R., fl. ?1957, *Chapman* 465 p.p. (FHO).
Mozambique. N: Cabo Delgado, Macondes, Mueda, fl. 19.x.1942, *Mendonça* 959 (C; LISC; LMA).
Also known from Kenya and Tanzania. In riverine forest.

In Kenya *A. tayloris* is a common species in forest and thicket near the coast, but outside this country its distribution is remarkably scattered. Only four non-Kenyan collections are known, including the three F.Z. specimens cited and a single one from SE. Tanzania: Newala, fl. 20.x.1959, *Hay* 73 (K). Both of the collections from Malawi are somewhat doubtfully localised. *Buchanan* 140 lacks any information but according to Exell (in F.Z. **1**: 27, 1960) is probably from the Shire Valley. *Chapman* 465 is a mixed collection (containing also *Landolphia buchananii* (Hallier f.) Stapf) and it is not certain that the locality notes apply to both species. Further collections of *A. tayloris* from Tanzania, Mozambique and Malawi would therefore be of great interest.

3. **Ancylobotrys petersiana** (Klotzsch) Pierre in Bull. Soc. Linn. Paris, N.S. **1**: 91 (1898). — Pichon in Mém. Inst. Fr. Afr. Noire **35**: 290, t. 19 fig. 1–3, map 20 (1953). — R. B. Drumm. in Kirkia **10**: 269 (1975). TAB. **96** fig. B. Type: Mozambique, Sena, *Peters* s.n. (B, †).
  Willughbeia petersiana Klotzsch in Peters, Reise Mossamb., Bot. **1**: 281 (1862). Type as above.
  Willughbeia senensis Klotzsch, tom. cit.: 282 (1862). Type: Mozambique, Sena, *Peters* s.n. (B†).
  Landolphia petersiana (Klotzsch) Dyer ex Hook. f. in Kew Gard. Rep. **1880**: 42 (1881); in Hook., Ic. Pl. **28**: t. 2756 (1903). — Stapf in F.T.A. **4**, 1: 47 (1902); in F.C. **4**, 1: 493 (1907). — Codd in Fl. Southern Afr. **26**: 260 (1963). Type as for *Ancylobothrys petersiana*.
  Landolphia monteiroi N.E. Br. in Monteiro, Delagoa Bay, 161, 163, 178 (1891). Type: Mozambique, Delagoa Bay, *Monteiro* 37 (K, holotype).
  Landolphia angustifolia K. Schum. ex Engl. in Abh. Preuss. Akad. Wiss. Berl. **1894**, Phys.-Math. Kl., **1**: 34 (1894). Type from Tanzania.

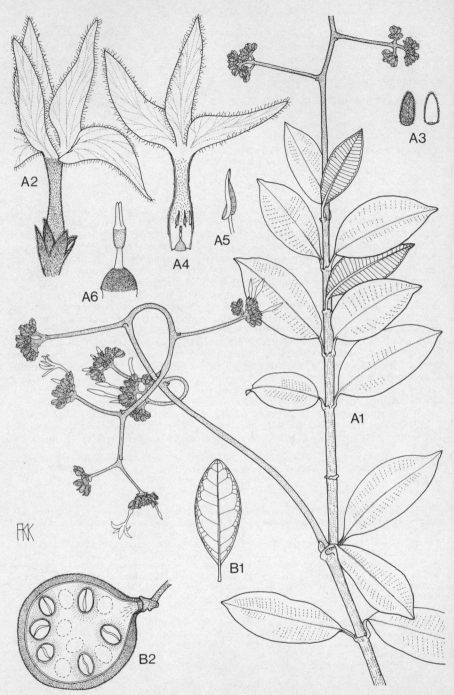

Tab. 96. A. — ANCYLOBOTRYS TAYLORIS. A1, habit ($\times\frac{2}{3}$); A2, flower ($\times\frac{8}{3}$); A3, calyx segment, dorsal (left) and ventral (right) views ($\times\frac{8}{3}$); A4, part of corolla opened out ($\times\frac{8}{3}$); A5, stamen, lateral view showing dorsal carina ($\times10$); A6, gynoecium ($\times10$), all from *Mendonça* 959. B. — ANCYLOBOTRYS PETERSIANA. B1, leaf, showing venation pattern ($\times\frac{2}{3}$), from *Andrada* 1917; B2, fruit, cut open to show seeds ($\times\frac{2}{3}$), from *Torre & Correia* 17032.

Landolphia senensis (Klotzsch) K. Schum. in Engl., Pflanzenw. Ost-Afr. **B**: 453 (1895). Type as for *Willughbeia senensis*.

Landolphia petersiana var. *rotundifolia* Dewèvre in Ann. Soc. Sci. Brux. **19**: 122 (1895). Type from Zanzibar.

Ancylobotrys rotundifolia (Dewèvre) Pierre in Bull. Soc. Linn. Paris, N.S. **1**: 92 (1898). Type as above.

Landolphia scandens (Schum.) Didr. var. *rotundifolia* (Dewèvre) Hallier f. in Jahrb. Hamb. Wiss. Anst. **17**, Beih. 3: 82 (1900). Type as above.

Landolphia scandens var. *petersiana* (Klotzsch) Hallier f., loc. cit. Type as for *Ancylobotrys petersiana*.

Landolphia scandens var. *stuhlmannii* Hallier f., tom. cit.: 83 (1900). Syntypes from Tanzania.

Landolphia scandens var. *angustifolia* (Engl.) Hallier f., tom. cit.: 84 (1900). Type as for *Landolphia angustifolia*.

Landolphia petersiana var. *angustifolia* (Engl.) Stapf in F.T.A. **4**, 1: 48 (1902). Type as above.

Landolphia petersiana var. *rufa* Stapf, loc. cit. Syntypes: Malawi, "Nyasaland", 1891, *Buchanan* 437 (BM; K); Mozambique, Tete, xii.1858, *Kirk* s.n. (K); Tete, ii.1859, *Kirk* s.n. (K).

Landolphia petersiana var. *tubeufii* Busse ex Stapf, op. cit.: 49 (1902). Type from Kenya.

Landolphia scandens var. *tubeufii* (Stapf) Busse in Engl., Bot. Jahrb. **32**: 171 (1902). Type as above.

Pacouria petersiana (Klotzsch) S. Moore in Journ. Linn. Soc., Bot. **37**: 180 (1905). Type as for *Ancylobotrys petersiana*.

Climbing shrub usually growing over trees and bushes, reaching 6 m. or more. Young branches greyish or rufous appressed-pubescent, tardily glabrescent, older twigs spotted with conspicuous round lenticels. Leaves thinly to thickly coriaceous, often strongly discolorous when dry, becoming blackish or dark green above and pale green or brown below; petiole 3–8 mm. long, pubescent or glabrous; lamina 5–11·5(13) × 1·8–5(6·5) cm., elliptic or obovate-elliptic, the apex rounded or obtuse, sometimes slightly cuspidate near the apex to form an indistinct acumen, the base rounded to subcordate. Upper leaf surface sparsely pubescent, veins ± level, obscure; lower surface completely glabrous or pubescent on midrib and sometimes also on lateral nerves, reticulation level, very conspicuous in dried leaves. Venation with lateral nerves well spaced, curving towards leaf apex. Inflorescences elongate panicles, each lateral branch terminating in a dense cymose flower-cluster, axes pubescent. Flowers very sweetly scented, white or creamy-white. Calyx 2–3 mm. long, lobes triangular or ovate, united at base, dorsally densely ferrugineous appressed-pubescent. Corolla tube 8–12 mm. long, slender, slightly inflated just below the middle at the level of the stamens, outer surface sparsely puberulent or glabrous, inner surface pubescent between the anthers and mouth; corolla lobes c. twice as long as tube, asymmetrically elliptic with acuminate apex, long-ciliate with delicate crispate white hairs. Stamens inserted 2–3 mm. from the base of the corolla tube, anthers subsessile, c. 1·5 mm. long. Ovary c. 1 mm. long, globose, with a dense ring of appressed pale hairs round the apex. Style, clavuncle and stigma 1·5–2 mm. long. Fruit c. 5 cm. in diameter, yellow or light orange when ripe, with pulpy flesh, edible and delicious.

Zimbabwe. N: Mutoko (Mtoko) Distr., Mutoko Tribal Trust Land, 5 km. upstream from Gurure Camp, fl. 3.xii.1968, *Müller & Burrows* 943 (K; SRGH). E: Mutare (Umtali) Distr., Commonage, 1100 m., fl. 30.x.1951, *Chase* 4161 (BM; BR; COI; K; LISC; PRE; SRGH). S: Ndanga Distr., Chitsa's village, st. 10.vi.1950, *Chase* 2410 (BM; K; LISC; SRGH). **Malawi.** S: Chikwawa Distr., Lengwe National Park, fl. xii.1970, *Hall-Martin* 1262 (SRGH). **Mozambique.** N: Angoche (António Enes), fl. 14.x.1965, *Gomes e Sousa* 4856 (COI; K; PRE). Z: 32 km. N. of Quelimane, fl. 10.viii.1962, *Wild* 5863 (BR; K; LISC; SRGH). T: Cahora Bassa, R. Mucangádzi, c. 570 m., fr. 29.i.1973, *Torre, Carvalho & Ladeira* 18912 (C; K; LISC; LMA; MO; WAG). MS: Sofala Prov. (Beira Distr.), N. of Macúti Beach, fl. 10.ix.1962, *Noel* 2491 (K; LISC; SRGH). GI: Inhambane Prov., Bazaruto I., Lighthouse hill, fl. & fr. 20.x.1958, *Mogg* 28451 (K; LISC; LMU; PRE; SRGH). M: Marracuene, Costa do Sol, fl. 10.viii.1959, *Barbosa & Lemos* in *Barbosa* 8668 (COI; K; LISC; SRGH).

Also known from S. Africa (Natal, Transvaal), Tanzania, Kenya and Madagascar. In dune scrub, among boulders on rocky hillsides, in mixed woodland.

I have seen no specimens of *A. petersiana* from either Ethiopia or Somalia, but Cufodontis (in Bull. Jard. Bot. Brux. **30**, Suppl.: 686, 1960) cites it from the former and Codd (in Fl. Southern Afr. **26**: 260, 1963) from the latter.

4. **Ancylobotrys amoena** Hua in Bull. Mus. Nation. Hist. Nat., Paris **5**: 186 (1899).—Pichon in Mém. Inst. Fr. Afr. Noire **35**: 274, t. 16 fig. 4 and t. 17 fig. 1–2, map 20 (1953). — F. White, F.F.N.R.: 346 (1962). — H. Huber in F.W.T.A., ed. 2, **2**: 60 (1963). — Fanshawe, Check List Woody Pl. Zambia: 4 (1973). Type from Guinea.

Landolphia scandens (Schum.) Didr. var. *ferruginea* Hallier f. in Jahrb. Hamb. Wiss. Anst. **17**, Beih. 3: 80 (1900). Syntypes from Nigeria and Tanzania.

Landolphia scandens var. *schweinfurthiana* Hallier f., tom. cit.: 81 (1900). Syntypes from the Sudan.

Landolphia scandens var. *rigida* Hallier f., loc. cit. Syntypes from Tanzania.

Landolphia amoena (Hua) Hua ex A. Chev. in Journ. Bot., Paris, **15**: 76 (1901). Type as for *Ancylobotrys amoena*.

Landolphia ferruginea (Hallier f.) Stapf in F.T.A. **4**, 1: 46 (1902). Syntypes as for *Landolphia scandens* var. *ferruginea*.

Landolphia petersiana var. *schweinfurthiana* (Hallier f.) Stapf, op. cit.: 48 (1902). Syntypes as for *Landolphia scandens* var. *schweinfurthiana*.

Pacouria petersiana (Klotzsch) S. Moore var. *schweinfurthiana* (Hallier f.) S. Moore in Journ. Linn. Soc., Bot. **37**: 180 (1905). Syntypes as above.

Pacouria amoena (Hua) Pichon in Mém. Mus. Nation. Hist. Nat., Paris, N.S. **24**: 144 (1948). Type as for *Ancylobotrys amoena*.

Landolphia nitida Lebrun & Taton in Expl. Parc Nation. Kagera, Miss. J. Lebrun 1937–38, fasc. 1: 105 (1948). Type from Zaire.

Scrambler or liane with stems persistently densely ferrugineous-velutinous. Leaves coriaceous, drying greyish-green to brown, somewhat paler below; petiole 9–12(15) mm. long, ferrugineous-pubescent; lamina 7–11·5 × 3·2–6·3 cm., oblong, ovate or even subcircular, the apex usually deflexed (so that the leaf cannot be pressed flat without folding), acute, rounded or with a short rounded acumen, the base rounded or rounded-cuspidate. Upper leaf surface mat or glossy, fairly densely to sparsely pubescent, with midrib impressed, other nerves level; lower surface mat, sparsely pubescent to subglabrous, midrib and lateral nerves raised, reticulation raised or level; venation with lateral nerves well spaced, curving towards leaf apex. Inflorescences elongate panicles, each lateral branch terminating in a dense cymose flower-cluster, axes densely ferrugineous-velutinous. Flowers white or cream with pale green or dull pink tube, sweet-scented. Calyx c. 4 mm. long, lobes narrowly triangular, united at the base, dorsally densely ferrugineous appressed-pubescent. Corolla tube 11–18 mm. long, slender, slightly inflated in the lower third, outer surface pubescent above the level of the calyx with indumentum continuing briefly onto each lobe in a narrow median line, inner surface glabrous; corolla lobes shorter than to equalling tube, narrowly elliptic with acute apex, ciliate with crispate white hairs. Stamens inserted 3–3·5 mm. from the base of the corolla tube; anthers subsessile, 1–1·5 mm. long. Ovary c. 1 mm. long, globose, glabrous or with a dense ring of hairs round the apex. Style, clavuncle and stigma 2–3 mm. long. Fruit up to c. 4 cm. in diameter, subglobose, orange. Seeds 8–12 mm. long.

Zambia. N: Mbala (Abercorn) Distr., Sunzu Mt. 40 km. from Mbala (Abercorn), 2400 m., fl. 28.viii.1960, *Richards* 13165 (BR; K; SRGH). W: Ndola Distr., fl. 22.vii.1954, *Fanshawe* 1389 (BR; K; LISC). **Malawi.** N: Nkhata Bay Distr., 16 km. S. of Nkhata Bay, 550 m., fr. 21.xii.1975, *Pawek* 10429 (K; SRGH).

Tropical Africa from Guinea eastwards to Zaire, Sudan, Rwanda, Uganda and Tanzania. In *Brachystegia* woodland, often on termitaria, on rough rocky ground and also in riparian vegetation.

Pichon (tom. cit.: 278, map 20) cited two *Kirk* specimens from Tete, Mozambique, under *A. amoena*, one of them being a syntype of *Landolphia petersiana* var. *rufa*. Having seen both the sheets concerned at K, I re-identify them as *A. petersiana*. No material of *A. amoena* is known from Mozambique. H. Huber, in F.W.T.A., ed. 2, **2**: 60 (1963), also cited *A. amoena* from Mozambique; this is presumably based on the same *Kirk* specimens.

The specimen "Rhodesia, 21.xi.1904, *Dunstan* s.n. (K)" which Pichon cited as *A. amoena*, tentatively from Zambia, is surely *A. petersiana*. Since the latter species does not extend into Zambia it may be concluded that the *Dunstan* specimen was collected in (Southern) Rhodesia (now Zimbabwe), although Exell & Hayes (in Kirkia **6**: 91, 1967) give only Zambia as the area worked by Dunstan.

My own survey of the flowers of *A. amoena* and *A. petersiana* suggests that they can be distinguished by both absolute and relative measurements: the calyx and corolla tube of *A. amoena* are longer than those of *A. petersiana*, while the corolla lobes of *A. amoena* are shorter, relative to the corolla tube, than in *A. petersiana*. These measurements are given in the present descriptions. However, Pichon (loc. cit.), who made a thorough investigation over the species' whole ranges, found wider variation in size and proportions of the flowers, making a greater overlap between the two species.

5. **Ancylobotrys pyriformis** Pierre in Bull. Soc. Linn. Paris, N.S. **1**: 127 (1899). — Pichon in Mém. Inst. Fr. Afr. Noire **35**: 280, t. 17 fig. 3–4, map 20 (1953). — H. Huber in F.W.T.A., ed. 2, **2**: 60 (1963). Type from Gabon.

    Ancylobotrys robusta Pierre, tom. cit.: 128 (1899). Type from Gabon.
    Landolphia robusta (Pierre) Stapf in F.T.A. **4**, 1: 43 (1902). Type as above.
    Landolphia pyriformis (Pierre) Stapf, op. cit.: 60 (1902). Type as for *Ancylobotrys pyriformis*.
    Pacouria pyriformis (Pierre) Pichon in Mém. Mus. Nation. Hist. Nat., Paris, N.S. **24**: 144 (1948). Type as above.
    Pacouria robusta (Pierre) Pichon, loc. cit. Type as for *Ancylobotrys robusta*.

Liane climbing to 7 m. or more. Young stems rufous or greyish puberulent. Leaves moderately coriaceous, drying glossy dark brown above and mat brown below, often with apex deflexed so that the leaf will not lie flat; petiole 18–25 mm. long; lamina 7–12 × 3–6 cm., ovate to oblong-elliptic, the apex acuminate or cuspidate-acuminate, the base rounded or acuminate. Upper leaf surface sparsely pubescent, midrib puberulent; all nerves impressed; lower leaf surface sparsely appressed-pubescent, especially on midrib; all nerves raised, the tertiary reticulation conspicuous; venation with lateral nerves well spaced, curved towards the leaf apex. Inflorescences few- to many-flowered capitate cymes arranged in long terminal panicles, with axes greyish-pubescent. Flowers white to yellow with pink or violet tube, sweet-scented. Calyx 1·2–2·5 mm. long, densely pubescent. Corolla tube 8–17 mm. long, slender, slightly inflated in the lower ⅓, outer surface pubescent above the level of the calyx, the indumentum sometimes continuing on to the corolla lobes; inner surface glabrous; corolla lobes much shorter than to slightly exceeding tube, elliptic, rounded or attenuate at apex, ciliate with crispate white hairs. Stamens inserted 1·5–2·5 mm. from base of corolla tube; anthers 1–1·5 mm. long. Ovary 0·5–1 mm. long, conical, puberulent. Style, clavuncle and stigma c. 1·5–2 mm. long. Fruit up to 12 cm. in diam., subglobose, orange; seeds 8–22 mm. long.

Zambia. W: Mwinilunga Distr., Zambezi R. c. 6·5 km. N. of Kalene Hill Mission, 0·8 km. above rapids, fl. 16.vi.1963, *Edwards* 793 (BR; K; SRGH).
    Gabon, Central African Rep., Congo Rep. and Zaire. In mushitu vegetation.

8. SABA (Pichon) Pichon

Saba (Pichon) Pichon in Mém. Inst. Fr. Afr. Noire **35**: 302 (1953).
Landolphia Sect. *Saba* Pichon in Mém. Mus. Nation. Hist. Nat., Paris, N.S. **24**: 140 (1948).

Tendrillous lianes. Latex present in all parts, abundant. Spines absent. Stipules absent. Cymes terminal, grouped into panicles with short to long peduncle. Calyx lobes imbricate, free to base. Corolla hypocrateriform; tube cylindric, slightly enlarged at level of anthers, pilose within especially towards the mouth; lobes contorted, overlapping to the left, ± elliptic, shorter or longer than tube. Stamens inserted below the middle of the corolla tube; anthers subsessile, without carina. Ovary hairy or glabrous (not in F.Z. area); style glabrous, clavuncle linear-fusiform, stigma not quite attaining level of anthers. Fruit glabrous, a many-seeded berry with sclerified pericarp.

Three species, all African, one extending to Madagascar and the Comores.

Saba comorensis (Bojer) Pichon in Mém. Inst. Fr. Afr. Noire **35**: 303, map 21 (1953). Neotype (of Pichon) from the Comores.

Vigorous liane climbing to 30 m.; stem terete, slash pink with white milky sap, bark brown, almost smooth. Young branches glabrous, lenticels conspicuous, small, circular, light-coloured, numerous. Leaves membranous, not strongly discolorous; petiole 8–14 mm. long, glabrous; lamina 7–19·5 × 4–9·5 cm., elliptic or ovate, rounded or obtuse at apex, rounded to subcordate at base; both surfaces glabrous, with lateral and tertiary nerves in relief; midrib raised below, impressed, level or slightly raised above; vein reticulation weakly scalariform. Tendrils common. Inflorescences terminal many-flowered paniculate cymes, forming large dense heads or lax and tendrillous; axes glabrous or tomentose. Flowers large, white, attractive, very sweet-scented. Calyx 2·5–3 mm. long, lobes strongly imbricate, free to base, oblong, rounded at apex, dorsally glabrous or tomentose, minutely ciliate. Corolla tube 18–24 mm. long, slender, slightly inflated just below the middle, outer surface completely glabrous or pubescent, especially towards the apex, inner surface pilose; corolla lobes 16–30 × 3–9 mm., oblong-elliptic or obovate-elliptic, rounded at apex, minutely and irregularly ciliate, lower surface glabrous or pubescent towards the base, upper surface glabrous or shortly pilose near the base. Stamens inserted just below the middle of the corolla tube; anthers c. 2·75 mm. long, subsessile. Ovary subglobose, c. 1 mm. long, with a ring of stiff erect hairs at the apex; style, clavuncle and stigma 4–5 mm. long. Fruit globose, up to 6 cm. in diameter when ripe, yellow or orange with white or orange pulp, edible.

Widespread throughout tropical Africa and found also in the Comores and Madagascar.

Inflorescence axes and dorsal surface of sepals glabrous; outer surface of corolla tube glabrous
- - - - - - - - - - - - var. *comorensis*
Inflorescence axes and dorsal surface of sepals tomentose; outer surface of corolla tube pubescent - - - - - - - - - - - - - - - -var. *florida*

These two varieties of *S. comorensis* cannot be distinguished in the vegetative state, as far as is known. Their distributions outside the F.Z. area are approximately sympatric, as Pichon's map shows, although only the glabrous variety is found on Madagascar and islands of the Comores.*

Var. **comorensis**. — Pichon, tom. cit.: 314 (1953). TAB. **97**.
    Vahea comorensis Bojer, Hort. Maurit.: 207 (1837). — A.DC. in DC., Prodr. **8**: 328 (1844). Type as for *Saba comorensis*.
    Willughbeia cordata Klotzsch in Peters, Reise Mossamb., Bot. **1**: 283 (1862). Type from the Comores.
    Landolphia florida Benth. var. *leiantha* Oliv. in Trans. Linn. Soc. **29**: 107 (1875). — Stapf in F.T.A. **4**, 1: 39 (1902). Type from Uganda.
    Landolphia comorensis (Bojer) K. Schum. in Engl., Bot. Jahrb. **15**: 406 (1893). Type as for *Saba comorensis*.

Zimbabwe. E: Mandidzudzure (Melsetter) Distr., between Lusitu and Haroni Rs., 370 m., fl. 24.iv.1962, *Wild* 5729 (COI; K; SRGH). S: Chiredzi Distr., near Sabi/Lundi Rs. junction, Gona-re-Zhou Game Reserve, fl. & fr. 31.v.1971, *Sherry* 305/71 (K; PRE; SRGH). **Malawi.** N: Rumphi Distr., St. Patrick's Seminary, Chelinda R., fl. 4.vii.1977, *Pawek* 12842 (MAL). S: Ntcheu Distr., Kapeni Stream near Ntoda mission, fl. 12.vi.1967, *Salubeni* 746 (K; MAL; SRGH). **Mozambique.** N: near foot of Ribáuè Mt., S. face, c. 610 m., fl. 19.vii.1962, *Leach & Schelpe* 11398 (K; LISC; SRGH). Z: Milanje, 42 km. on road to Quelimane, c. 700 m., fr. 2.xi.1967, *Torre & Correia* 15865 (LISC; LMA). MS: Vila Machado, R. Muda, fl. 28.ii.1948, *Mendonça* 3815 (C; LISC; LMU; WAG).
    In riverine vegetation and in open woodland.

Var. **florida** (Benth.) Pichon, tom. cit.: 309 (1953). Type from Nigeria.
    Landolphia florida Benth. in Hook., Niger Fl.: 444 (1849). — Stapf, op. cit.: 38 (1902). Type as above.
    Vahea florida (Benth.) F. Muell. in Wittst., Org. Const. Pl.: 258 (1878). Type as above.
    Landolphia comorensis var. *florida* (Benth.) K. Schum., tom. cit.: 404, fig. 1B & 2 (1893). Type as above.
    Pacouria florida (Benth.) Hiern, Cat. Afr. Pl. Welw. **1**: 662 (1898). Type as above.
    Saba florida (Benth.) Bullock in Kew Bull. **13**: 391 (1959). — H. Huber in F.W.T.A., ed. 2, **2**: 61 (1963). — Fanshawe, Check List Woody Pl. Zambia: 38 (1973). — R. B. Drumm. in Kirkia **10**: 269 (1975). Type as above.

Zambia. N: Nsama Distr., fl. 12.ix.1958, *Fanshawe* 4809 (BR; FHO; K; LISC; SRGH). **Malawi.** N: Nkhata Bay Distr., Chombe Estate, fl. 21.vi. 1960, *Adlard* 374 (FHO; K; PRE; SRGH). C: Dedza Distr., Chongoni Forest School, fl. 14.iii.1963, *Salubeni* 7 (SRGH). S: Zomba and E. end of L. Chilwa (Shirwa), fl. & fr. x.1861, *Meller* s.n. (K). **Mozambique.** N: Mandimba Distr., 1·6 km. E. of Mandimba border post, fl. 2.v.1960, *Leach & Brunton* 9900 (K; LISC; SRGH). Z: Lugela, road from Mocuba to Mobede, near R. Munhamade, fl. 23.v.1949, *Andrada* 1477 (COI; LISC). MS: Mwangezi R., Lower Búzi, 150 m., fl. 25.xi.1906, *Swynnerton* 1081 (BM).
    Habitat as for var. *comorensis*.

* Editor's note: Recent studies by Leeuwenberg et al., based on large numbers of specimens, seem to suggest that these varieties perhaps cannot be maintained.

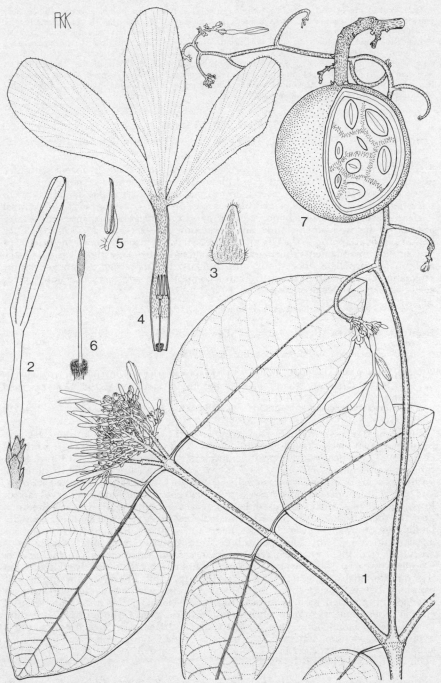

Tab. 97. SABA COMORENSIS var. COMORENSIS. 1, habit (×⅔), from *Garcia* 931; 2, flower bud
(×2); 3, calyx segment, dorsal view (×6); 4, part of flower opened out (×2); 5, stamen
(×6); 6, gynoecium (×6), 2–6 from *Torre & Paiva* 10615; 7, fruit (×⅔), from *Torre &*
Correia 15865.

Bullock, loc. cit. (1959), stated that *Vahea comorensis* is a *nomen nudum* and that the earliest validly published name for this species is *Landolphia florida*; he therefore made the combination *Saba florida* (Benth.) Bullock which has since been widely used. However, a fresh look at the original publication of *Vahea comorensis* shows that Bojer described his plant as a member of the *Apocynaceae*, a woody liane flowering in September and bearing fruits the shape and colour of an orange, which grows in the mountains around Musamodo, the capital of Anjouan I. This is surely adequate to validate the name. Moreover, even if *Vahea comorensis* Bojer had been a *nom. nud.* it would have been validated by A. De Candolle (loc. cit., 1844) prior to the publication of *Landolphia florida* in 1849.

9. HUNTERIA Roxb.

Hunteria Roxb., Fl. Ind., ed. 2, **1**: 695 (1832). — Pichon in Bol. Soc. Brot., Sér. 2, **27**: 88 (1953).

Trees, shrubs and lianes. Latex present. Spines and tendrils absent. Stipules absent. Stems and leaves completely glabrous. Inflorescences cymose, terminal, rarely also axillary; pedicels present, up to 1 cm. long. Calyx glabrous; lobes free to base, herbaceous, imbricate, not revolute; scales present, covering part of the ventral surface of sepals. Corolla hypocrateriform, externally glabrous, internally hairy above and below stamens; tube not developing splits (as in *Pleiocarpa*); lobes contorted, overlapping to the left, not ciliate. Stamens inserted above the middle of the corolla tube; filaments short, anthers dorsifixed, introrse, without carina. Ovary glabrous, 2-carpellate, apocarpous, with 2–30 ovules per carpel; clavuncle present; stigmatic apiculus well developed. Fruit a compound berry; mericarps not beaked; seeds with smooth testa; endosperm horny.

A genus of 10 species with one in S.E. Asia and E. Africa and the rest in W. and C. Africa.

Hunteria zeylanica (Retz.) Gardn. ex Thw., Enum. Pl. Zeyl.: 191 (1860). Type the description.
   Cameraria zeylanica Retz., Obs. Bot. **4**: 24 (1786). Type as above.

Var. **africana** (K. Schum.) Pichon in Bol. Soc. Brot., Sér. 2, **27**: 106, t. 2 fig. 6, map 13 (1953). — Dale & Greenway, Kenya Trees & Shrubs: 47 (1961). TAB. **98**. Syntypes from Tanzania.
   Hunteria africana K. Schum. in Engl., Pflanzenw. Ost-Afr. **C**: 317 (1895). — Stapf in F.T.A. **4**, 1: 104 (1902). Syntypes as above.

Shrub or tree 1–40 m. high; bark grey, smooth, with cream-coloured slash and bright orange underbark, yielding milky latex. Young twigs glabrous. Leaves thinly coriaceous, ± concolorous, opposite, completely glabrous; petiole 7–13 mm. long, not winged; lamina (4·7)7–22 × (1·3)2–4 cm., oblong to elliptic or slightly obovate, with apex acute or obtuse or occasionally acuminate to a short round-tipped acumen, and base acute; upper surface glossy, with midrib channelled and other nerves raised; lower surface mat, all nerves raised; lateral nerves 3–7 mm. apart, ± parallel, joined near the margin in a neatly looped vein. Inflorescences terminal (and occasionally lateral) few- to many-flowered cymes, open and regularly branched, with glabrous axes; peduncle 3–25 mm. long, pedicels 2–3 mm. Calyx c. 1 mm. long, lobes oblong-apiculate, free, glabrous; each sepal bearing a small scale on its ventral surface. Corolla white or pale yellow; tube 7–8 mm. long, narrow, very sightly contracted in the middle, glabrous externally, sparsely pubescent internally in upper half especially below the stamens; lobes c. 6 × 2–2·5 mm., narrowly ovate, with strongly overlapping auricled base, glabrous. Stamens inserted 5·5–6 mm. from base of corolla tube; filaments minute, anthers 0·75–1 mm. long. Gynoecium 5·5–6·5 mm. long; ovary 0·5 mm. long, conical, glabrous, of 2 free carpels each containing 2 ovules; style slender, flattened; clavuncle 2-lobed, stigma apiculate. Fruit of 2 subglobose mericarps each 11–27 × 6–21 mm., 2-seeded, at first green, becoming yellow to orange. Seeds 1–1·5 mm. long.

Mozambique. Z: Mopeia, st. 24.viii.1972, *Bowbrick* JC10 (LISC). MS: 25 km. from Lacerdonia, N. of new railway, 200 m., fl. buds 6.xii.1971, *Müller & Pope* 1926 (LISC; PRE; SRGH).
   Also known from Kenya and Tanzania (var. *africana*). In thicket and mixed evergreen forest.

Pichon (loc. cit.) has divided *Hunteria zeylanica* into three varieties, as follows. Var. *zeylanica* has leaves almost always over 2·5 cm. wide, and with a well-defined acumen; it occurs

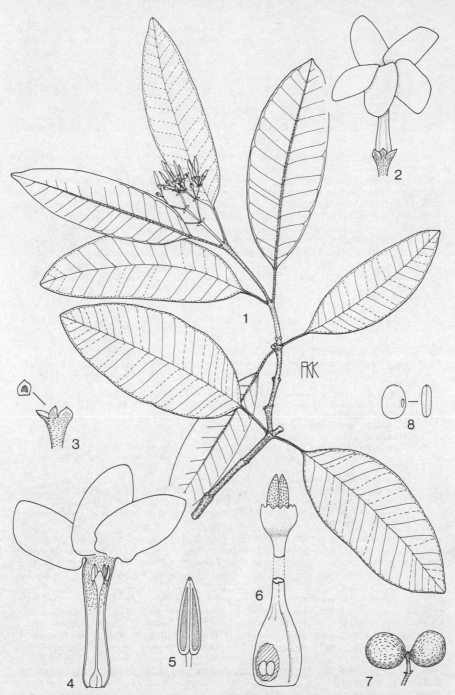

Tab. 98. HUNTERIA ZEYLANICA var. AFRICANA. 1, habit (×⅔), from *Andrada* 1062; 2, flower
(×⅚); 3, part of calyx, with detail showing scale on ventral surface of sepal (× 4); 4, part of
corolla opened out (× 4); 5, stamen (× 20); 6, apex and base of gynoecium, the latter with
wall of one loculus removed to show the two ovules (× 20), 2–6 from *Tinley* 2369; 7, fruit
(×⅔); 8, seed, ventral and lateral views (× 1), 7–8 from *Müller & Pope* 1865.

in India, Sri Lanka, Burma, Thailand, Viet-Nam, Kampuchea, Malaya and Sumatra. Var. *salicifolia* (Wall. ex A. DC.) Pichon has leaves rarely more than 2·3 cm. wide, and without a distinct acumen; it occurs in India and Sri Lanka. The African variety, var. *africana*, has leaves all or mostly over 2·5 cm. wide and without a distinct acumen.

10. PLEIOCARPA Benth.

Pleiocarpa Benth. in Benth. & Hook., Gen. Pl. **2**: 699 (1876). — Pichon in Bol. Soc. Brot., Sér. 2, **27**: 123 (1953).

Trees, shrubs and lianes. Latex present. Spines and tendrils absent. Stipules absent. Stems and leaves completely glabrous. Leaves opposite or verticillate. Inflorescences cymose, axillary, rarely also terminal; pedicels short or absent. Calyx glabrous; lobes free to base, herbaceous, imbricate, not revolute, without scales on inner surface. Corolla hypocrateriform, externally glabrous, internally pubescent below the stamens; tube often with 5 splits developing at level of stamens; lobes contorted, overlapping to the left, not ciliate. Stamens inserted near mouth of corolla tube; filaments short; anthers dorsifixed, introrse, without carina. Ovary glabrous, 2–5-carpellate, apocarpous, with 1–4 ovules per carpel; clavuncle present; stigma reduced to a sessile area or rarely represented by a rudimentary apiculus. Fruit a compound berry; mericarps not beaked. Seeds with smooth testa; endosperm horny.

An African genus of 7 species.

Pleiocarpa pycnantha (K. Schum.) Stapf in Dyer, F.T.A. **4**, 1: 99 (1902). — Eggeling & Dale, Indig. Trees Uganda Prot., ed. 2: 29 (1951). — Pichon in Bol. Soc. Brot., Sér. 2, **27**: 128, t. 4 fig. 1 (1953). — F. White, F.F.N.R.: 350 (1962). — R. B. Drumm. in Kirkia **10**: 269 (1975). TAB. **99**. Type from Uganda.
  Hunteria pycnantha K. Schum. in Engl., Bot. Jahrb. **23**: 222 (1896). Type as above.
  Pleiocarpa welwitschii Stapf ex Hiern, Cat. Afr. Pl. Welw. **1**: 665 (1898). Type from Angola.
  Pleiocarpa tubicina Stapf in Kew Bull. **1898**: 304 (1898). Type from Zaire.
  Hunteria breviloba Hallier f. in Jahrb. Hamb. Wiss. Anst. **17**, Beih. 3: 189 (1900). Type from the Congo Rep.
  Pleiocarpa micrantha Stapf in Dyer, F.T.A. **4**, 1: 100 (1902). — Irvine, Woody Pl. Ghana: 630 (1961). Type from Ghana.
  Pleiocarpa flavescens Stapf, tom. cit.: 101 (1902). Type from Ghana.
  Pleiocarpa breviloba (Hallier f.) Stapf, tom. cit.: 102 (1902). Type as for *Hunteria breviloba*.
  Pleiocarpa microcarpa Stapf, loc. cit. Type from Zaire.
  Pleiocarpa bagshawei S. Moore in Journ. Bot. Lond. **45**: 49 (1907). Type from Uganda.
  Pleiocarpa swynnertonii S. Moore in Journ. Linn. Soc., Bot. **40**: 138 (1911). Type: Zimbabwe, Chirinda forest, 1130–1220 m., fl. 27.ix.1905, *Swynnerton* 14 (BM, lectotype; SRGH, isolectotype).

An evergreen shrub or tree 1·5–30 m. tall; bark grey, smooth or slightly scaly; slash whitish, cream or pale brown with darker yellow striations, brownish yellow beneath, yielding milky latex; branches subsarmentose. Young twigs glabrous, angled, sometimes with verticillate branching-pattern. Leaves thinly coriaceous, opposite or in whorls of 3–5, completely glabrous; petiole 3–20 mm. long, angled, often narrowly winged almost to the base by a continuation of the lamina; lamina (4)6–13·5 × (1·3)2·2–5·5 cm., elliptic to obovate or narrowly obovate, the apex cuspidate with short rounded acumen or acute with rounded tip, the base attenuate; upper surface glossy, with all nerves raised (midrib both prominent and channelled); lower surface paler and mat, all nerves raised; lateral nerves ± parallel, 2–4 mm. apart, joining near the margin in a neatly looped vein; edge of leaf forming little pleats when pressed flat. Flowers white, fragrant, in dense axillary fascicles on new and old wood; pedicels up to 1 mm. long, glabrous. Calyx 1·5–2 mm. long; sepals ovate, free, glabrous. Corolla tube 5–8 mm. long, narrowed in the middle, glabrous externally, pubescent internally in upper half, hairs longest just below the anthers; corolla lobes 2–4 × 1·5–2·5 mm., elliptic, glabrous, not strongly overlapping. Stamens inserted 4–6 mm. above base of corolla tube; filaments minute, anthers 0·5–1 mm. long. Gynoecium 4–6·5 mm. long; ovary c. 1 mm. long, cylindrical, glabrous, of 2 free carpels each containing 1–4 ovules; style slender, flattened; clavuncle 2-lobed; stigma reduced to a flat region at apex of clavuncle. Fruit

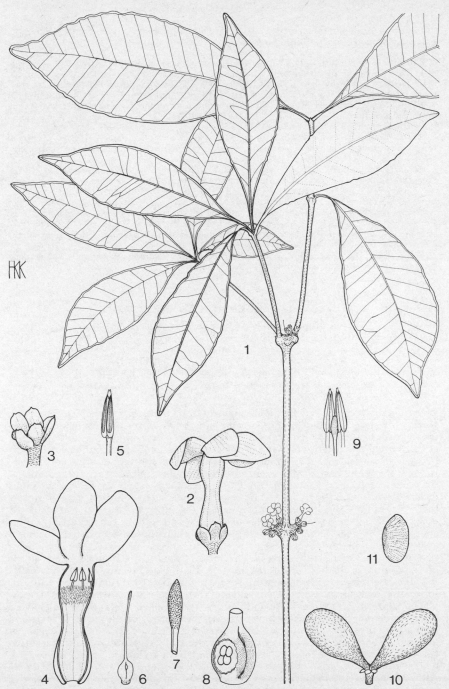

Tab. 99. PLEIOCARPA PYCNANTHA. 1, habit ($\times \frac{2}{3}$), from *Gomes e Sousa* 4407; 2, flower ($\times \frac{2}{3}$); 3, calyx ($\times 4$); 4, part of corolla opened out ($\times 4$); 5, stamen ($\times 12$); 6, gynoecium ($\times 4$); 7, enlargement of 6 showing papillose clavuncle ($\times 24$); 8, diagrammatic enlargement of ovary with wall of one carpel partly removed to show ovules attached to ventral wall; 9, relative positions of anthers and clavuncle at anthesis ($\times 12$), from *Goldsmith* 38/65; 10, fruit ($\times 1$); 11, seed ($\times 1$), 10–11 from *Goldsmith* 144/67.

comprising 2 clavate mericarps 7–20 × 4·5–18 mm., pale green to orange when ripe, each containing 1 or 2 salmon-pink seeds.

Zambia. N: north bank of R. Luapula near Chembe Ferry, st. 5.x.1947, *Brenan & Greenway* 8011 (BR; FHO; K). **Zimbabwe.** E: Chipinge (Chipinga) Distr., Chirinda forest, 1100 m., fl. x.1965, *Goldsmith* 38/65 (BR; K; LISC; PRE; SRGH). **Mozambique.** Z: Maganja da Costa, Gobene forest, 50 km. from Vila de Maganja, c. 20 m., fl. 12.xi.1966, *Torre & Correia* 14541 (C; LISC; LMA; MO; SRGH; WAG). MS: 25 km. from Lacerdonia, N. of new railway, 200 m., fl. 6.xii.1971, *Müller & Pope* 1916 (K; LISC; PRE; SRGH). GI: Gaza, Macia, S. Martinho do Bilene, fl. & fr. 4.iv.1969, *Balsinhas* 1453 (LISC).

Widely distributed throughout tropical Africa. In a variety of habitats (mixed evergreen forest, deciduous forest, woodland on dunes, *Brachystegia* thicket) but usually on sandy soil.

Pichon, loc. cit., recognised two varieties within this species, var. *pycnantha* and var. *tubicina* (Stapf) Pichon. They were said to differ in the size of corolla lobes, var. *pycnantha* having lobes 2·4–5·6 × 1·4–3·45 mm. and var. *tubicina* 0·75–2·5(3·1) × 0·75–2 mm. In var. *pycnantha*, however, a branch usually was found to bear a mixture of long- and short-lobed flowers. Var. *tubicina* was believed to occur throughout the range of the species, var. *pycnantha* in E. Africa. Pichon saw only one specimen of *P. pycnantha* from the F.Z. area, *Swynnerton* 14. All the F.Z. specimens I have seen with expanded flowers belong to var. *pycnantha*, but many of the collections lack such flowers and cannot be assigned to variety.

According to Pichon (tom cit.: 131), the carpels of *P. pycnantha* are generally 1–2-ovulate, but I have found 4-ovulate examples, as TAB. **99** shows.

11. VOACANGA Thouars

Voacanga Thouars, Gen. Nov. Madag.: 10 (1806).
Orchipeda Bl., Bijdrage 1026 (1826). — Miquel, Ann. Mus. Lugd.-Bat. **1**: 316 (1864).
Annularia Hochst. in Flora **24**: 670 (1841).
Cyclostigma Hochst. ex Endl., Gen. Pl., Suppl. **2**: 56 (1842); in Flora **27**: 828 (1844).
Piptolaena Harv. in Hooker, Journ. Bot., Lond. **1**: 25 (1842).
Pootia Miq., Versl. en Mem. Kon. Akad. Wetsch. **6**: 192 (1857).

Shrubs or trees, repeatedly dichotomously branched and with bark usually with some white latex; 2 inflorescences in the forks when flowering. Branchlets terete, mostly deeply sulcate when dry, with white latex. Leaves opposite, those of a pair equal or unequal, usually petiolate; petioles or leaf bases of a pair connate into a short ocrea, with a single row of colleters in the axils; lamina elliptic or obovate or narrowly so, cuneate at the base or decurrent into the petiole. Inflorescence in the forks of branches usually long-pedunculate, corymbose, usually rather lax. Bracts deciduous or less often (not in the F.Z. area) more or less persistent. Flowers actinomorphic, often fragrant. Calyx green, campanulate to cylindrical, usually shed with the corolla, with colleters inside. Corolla mostly creamy or yellow; tube shorter or (in the species of the F.Z. area) only slightly longer than the calyx, twisted, variously shaped; lobes in bud overlapping to the left, spreading or recurved. Stamens exserted or included; anthers sessile, narrowly triangular, acuminate at the sterile apex, sagittate at the base, glabrous, introrse. Pistil glabrous; ovary mostly broadly ovoid; carpels free or connate at the base, surrounded by a ring-shaped entire or lobed disk; disk adnate to the abaxial sides of the carpels; style thickened at the apex; clavuncula with a thin, recurved, undulate, entire, or lobed ring at the base, obovoid with 5 short lateral lobes, coherent with the connectives of the anthers. Style and stigma shed together with the corolla; stigma short. Fruit two free or less often partly or completely united carpels (the latter not in the F.Z. area); wall soft, carnose, often thickish, creamy or yellow. Seeds surrounded by a yellow or orange pulpy aril, usually numerous, large, obliquely ellipsoid or less often reniform, at the hilar side with one deep groove to half their width, less deeply grooved at the other sides, with a minutely tuberculate or honeycomb-like structure; endosperm copious, starchy, creamy to white, ruminate, surrounding the spathulate embryo.

An Old World genus of 12 species, 7 of which occur in Africa.

Leaves acuminate, sometimes acute or obtuse, herbaceous when fresh, membranous or papery when dry; calyx not clasping the corolla and not shed with it, with mostly partly recurved lobes which are 0·8–1·3 times as long as the tube; corolla lobes obovate, narrowly obovate or narrowly elliptic. In dry soil - - - - - - - - - -1. *africana*

Leaves rounded or obtuse, often thickly coriaceous; calyx clasping the corolla and shed with it, barrel-shaped to cylindrical, 2·5–4 times as long as the lobes; corolla lobes broadly obcordate. In swamps - - - - - - - - - 2. *thouarsii*

1. **Voacanga africana** Stapf in Journ. Linn. Soc. **30**: 87 (1894); in F.T.A. **4**, 1: 157 (1902). — H. Huber in F.W.T.A., ed. 2, **2**: 67 (1963). TAB. **100**. Type from Sierra Leone.

V. schweinfurthii Stapf in Kew Bull. **1894**: 21 (1894); in F.T.A. **4**, 1: 155 (1902). Type from Zaire.

V. boehmii K. Schum. in Engler, Pflanzenw. Ost-Afr. **C**: 317 (1895); in Engler & Prantl, Nat Pflanzenfam. **4**, 2: 149 (1895). — Stapf in tom. cit.: 16 (1894). Type from Tanzania.

V. glabra K. Schum. in Engler & Prantl, Nat. Pflanzenfam. **4**, 2: 149 (1895). Type from Cameroon.

V. puberula K. Schum. in Engler & Prantl, loc. cit. — Stapf, tom. cit.: 156 (1902). Type from Gabon.

V. schweinfurthii var. *parviflora* K. Schum. in Engler, Bot. Jahrb. **23**: 226 (1896). Type from Togo.

V. angolensis Stapf ex Hiern, Cat. Welw. Afr. Pl. **1**: 668 (1898). — Stapf, tom. cit.: 154 (1902). Type from Angola.

V. spectabilis Stapf, tom. cit.: 155 (1902). Type from Angola.

V. lutescens Stapf, tom. cit.: 157 (1902). Type: Mozambique, Lower Zambezi R., between Lupata and Sena, *Kirk* 31 (K, lectotype).

V. magnifolia Wernham in Cat. Talbot's Nigerian Pl.: 62 (1913). Type from Nigeria.

V. eketensis Wernham in Journ. Bot., London **52**: 25 (1914). Type from Nigeria.

V. glaberrima Wernham, loc. cit. Type from Nigeria.

V. bequaerti De Wild., Pl. Bequaert. **1**: 402 (1922). Type from Zaire.

V. lemosii Philipson in Exell, Cat. Vasc. Pl. S. Tomé: 241 (1944). Type from S. Tomé.

V. africana var. *glabra* (K. Schum.) Pichon, Bull. Mus. Nation. Hist. Nat., Paris, Sér. 2, **19**: 412 (1947). Type as for *V. glabra*.

V. africana var. *auriculata* Pichon, loc. cit. Type from Tanzania.

V. africana var. *lutescens* (Stapf) Pichon, loc. cit. Type as for *V. lutescens*.

V. schweinfurthii var. *puberula* (K. Schum.) Pichon, tom. cit.: 413 (1947), partly, excl. of syn. *V. diplochlamys*.

Shrub-like tree or shrub 1–10(25) m. high. Trunk terete, 2–30 (40) cm. in diam., bark pale grey brown, smooth or shallowly fissured near the base with some white latex. Branches lenticellate; branchlets glabrous, puberulous, or pubescent, with more latex than in the bark. Leaves shortly petiolate or sessile; petiole glabrous to pubescent, up to 2 cm. long, those of a pair connate into a short ocrea, which is not widened into intrapetiolar stipules, with a single row of colleters in the axils; lamina usually pale green when plant in flower, when in fruit usually darker, especially above, herbaceous when fresh, membranous to papery when dry, variable in shape and size, elliptic or narrowly elliptic, 1·5–3·5(4) times as long as wide, 7–41·5 × 3–20 cm. (sometimes a little smaller), cuneate or decurrent into the petiole or base, not connate-perfoliate when sessile, acuminate with an often blunt apex or less often acute or obtuse in some leaves, glabrous on both surfaces to pubescent beneath all over and on the midrib above, with 8–22 secondary veins on each side; tertiary venation inconspicuous. Inflorescences usually long-pedunculate, 6–25 × 4–15 cm., usually many-flowered, fairly lax, 3·5–15·5 cm. long (incl. peduncle); peduncle pale green and glabrous to sparsely pubescent as branches and pedicels, usually slender; pedicels 3–20 mm. long. Bracts deciduous, usually all shed before the first bud reaches full size, about as long as the calyx, ovate, obtuse, with a few persistent colleters at the edges of the axils; upper bracts often narrower; all bracts leaving conspicuous scars. Flowers malodorous. Calyx pale green (paler inside), 7–19 mm. long (when lobes erect), deciduous after the corolla has been shed and before the fruit develops, glabrous or puberulous, out and inside, inside with a zone of colleters from 1–2 mm. above the base of the tube to 1–2 mm. below the base of the lobes (the uppermost colleter near the edge of the base of the lobes and the remainder are arranged irregularly or in 1–3 rows above each other); tube cyathiform, 3·5–9 mm. long; lobes erect, 0·8–1·3 times as long as the tube, subequal, broadly ovate to oblong, 0·7–1·3(1·7) times as long as wide, 3·5–8 × 3·5–8 mm., obtuse, rounded, truncate, retuse, or emarginate at the apex, entire, imbricate in bud, usually partly recurved. Corolla creamy, greenish-creamy, yellow, or less often white, in the mature bud 17–31 mm. long, incl. the lobes (the lobes ½–⅔ the length of the bud, being 8·5–19 mm., forming an almost conical head with a blunt apex), glabrous or less often minutely puberulous on both sides and inside often pubescent from about 5 mm. above base to the insertion of the stamens, tube slightly shorter to slightly longer than the calyx (when the lobes erect), 7–15 mm. long, almost cylindrical, twisted from 2–3 mm. above the base, contracted at the base, in the middle, and at the mouth, at the mouth 3–5 mm. wide; lobes twisted in bud, obovate, narrowly obovate, or elliptic, 1·4–2·5 times as long as the tube, (1·1)1·5–2·2 times as long as wide, 12–37 × 7–16 mm., rounded or obtuse and usually with an upcurved margin at the apex, entire, spreading and often recurved later. Stamens exserted for 0·5–1·2 mm. or occasionally just included, inserted 2–3 mm. below the corolla-mouth; anthers sessile, narrowly triangular, 4–5 × 1·3–2·5 mm., acuminate at the sterile apex and sagittate at the base, glabrous, usually twisted with the corolla. Pistil glabrous, 7–12·5 mm. long (when style straightened out); ovary 1·6–2·5 × 1·7–3 × 1·2–2 mm., of two carpels. Carpels connected at the base only by an entire ring-shaped disk-like 0·8–1·2 mm. high incrassation and at the apex by the style; style split at the base, narrowly obconical, 4–8 mm. long, twisted and curled at the base, at the apex slightly narrower than the clavuncula; clavulcula 1–1·7 × 1–1·7 mm., with a ring of 1·7–3 mm. in diameter. Ovules about 200 in each carpel. Fruit composed of two separate mericarps of which often only one develops; carpels dark and very pale green-spotted, obliquely subglobose, often slightly wider than long and laterally compressed, 3–8 × 3–8 × 2·5–7 cm., 2-valved; wall 5–15 mm. thick, creamy inside and on section. Seeds many, dark brown, dull, with orange aril, obliquely ellipsoid, 7–10 × 3·5–5 × 3–4 mm., laterally with 4–5 grooves, rough, minutely tuberculate.

Zambia. Z: Kabompo Distr., Kabompo R., below Boma, fl. 23.xi.1952, *Holmes* 1010 (FHO; K; SRGH). N: Mansa (Fort Rosebery) Distr., Lake Bangweulu shore, Samfya Mission, fr. 20.viii.1952, *White* 3094 (BM; BR; FHO; K; MO). E: Lundazi, fl. 12.x.1961, *Grout* 265 (FHO). **Zimbabwe.** E: Chimanimani Distr., near S. bank of Haroni R., 340 m. fl. 29.xi.1983, *Müller* 3775 (K; SRGH). **Malawi.** N: Karonga Distr., Kaporo, fl. 2.i.1974, *Pawek* 7730 (K; SRGH; UC). S: Shire Highlands, Blantyre, summit of the mountain, fl. 6.vii.1879, *Buchanan* 9 (E; K). **Mozambique.** N: Entre Rios, 700 m., fl. xii.1931, *Gomes e Sousa* 839 (LISC). Z: 50 km. NE of Mopeia Velha, on road to Quelimane, fl. 7.xii.1971, *Müller & Pope* 1952 (K; LD; LISC; SRGH). MS: Búzi R., below Espungabera (Spungabera), 600 m., fl. i.1962, *Goldsmith* 5/62 (K; LISC; MO; SRGH).

Widespread in tropical Africa on the continent and on the islands in the Gulf of Guinea. Open woodland or light forest, riverine forests, in savannas only in moist places; 0–1000 m.

2. **Voacanga thouarsii** Roem. & Schult., Syst. **4**: 439 (1819). — Stapf in F.T.A. **4**, 1: 154 (1902). — Codd in Fl. Southern Afr. **26**: 273 (1963). — H. Huber in F.W.T.A., ed. 2, **2**: 67 (1963). — Markgraf in Fl. Madag., Apocynac.: 219 (1976). Type from Madagascar.

V. dregei E. Mey., Comm. Pl. Afr. Austr. 189 (1838). Type from S. Africa.

Annularia natalensis Hochst. in Flora **24**: 670 (1841). Type from S. Africa.

Cyclostigma natalense (Hochst.) Hochst. ex Endl., Gen. Pl. Suppl. **2**: 56 (1842); in Flora **27**: 828 (1844). Type as for *Annularia natalensis*.

Piptolaena dregei (E. Mey.) A. DC., Prod. **8**: 358 (1844). Type from S. Africa.

Voacanga obtusa K. Schum. in Engler & Prantl, Nat. Pflanzenfam. **4**, 2: 149 (1895). — Stapf, tom. cit. 153 (1902). Type from Zaire.

V. thouarsii var. *dregei* (E. Mey.) Pichon in Bull. Mus. Nation. Hist. Nat., Paris, Sér. 2, **19**: 415 (1947). Type as for *V. dregei*.

V. thouarsii var. *obtusa* (K. Schum.) Pichon, loc. cit. Type as for *V. obtusa*.

Usually a small tree, 2–15(20) m. high; trunk terete, 4–40(80) cm. in diam., bark pale grey-brown, smooth, with small lenticels, with some white latex. Branches with or without some large lenticels; branchlets glabrous or minutely puberulous, with much latex. Leaves shortly petiolate; petiole 8–25 mm. long, glabrous or minutely puberulous at the base, those of a pair connate into a short ocrea (widened into intrapetiolar stipules), with a single row of colleters in the axils; lamina often slightly glaucous with the costa often pale, often thickly coriaceous (also when fresh), narrowly obovate or less often narrowly elliptic, 2–4 times as long as wide, 6–25 × 2–9 cm., obtuse or rounded, cuneate or decurrent into the petiole, glabrous and usually with numerous minute pits on both surfaces; secondary veins 12–20 on each side, inconspicuous in fresh leaves; tertiary venation inconspicuous. Inflorescences long-pedunculate, fairly lax, without peduncle 4–7 × 4–7 cm., few-flowered, monochasial or nearly so. Peduncle 5–14 cm. long, stout, glabrous; pedicels 8–15 mm. long except for the thickened 2–5 mm. long apex, often mintuely puberulous above. Bracts dediduous and leaving conspicuous scars, ovate, rounded at the apex, with a single row of large colleters in the axils; lower up to 10 × 7 mm., other smaller. Flowers sweet-scented. Calyx pale, to dark green, with hyaline margin, fleshy, clasping the corolla tube and shed together with the corolla, from

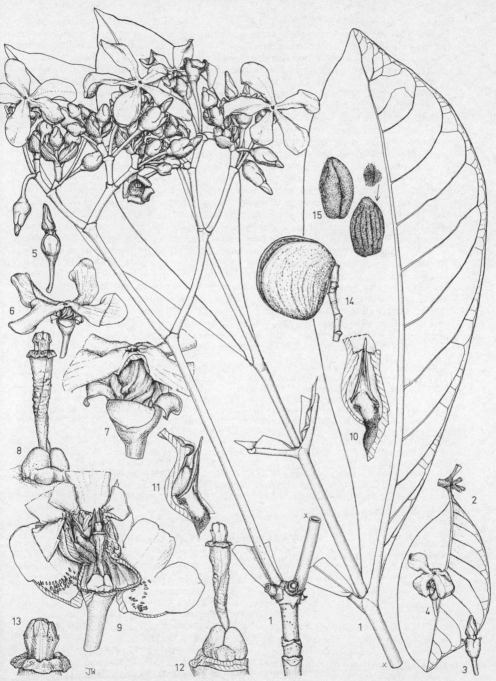

Tab. 100. VOACANGA AFRICANA. 1, flowering branches (× ½), from *Leeuwenberg* 9485; 2, leaf
(× ½) from *Leeuwenberg* 11927; 3 and 5, flower buds (both × ½!), 3 from *Leeuwenberg* 11927
and 5 from *Leeuwenberg* 9256; 4 and 6, flowers (both × ½!); 7, flower of which limb partly
removed (× 1½); 8, pistil (× 3), 7–8 from *Leeuwenberg* 9256; 9, flower, opened up (× 1½); 10,
11 stamens (× 3); 12, pistil (× 3); 13, pistil head (× 6), 9–13 from *Leeuwenberg* 9485; 14,
open carpel (× ¼); 15, seeds (× 1½), 14–15 from *Leeuwenberg* 8660.

abscission layer 13–25 mm. long, 7–13 mm. wide, outside minutely papilose except for the glabrous apices of the lobes, inside with several irregular rows of colleters in the lower half of the tube; tube barrel-shaped to cylindrical, 2·5–4 times as long as the lobes, 10–16 mm. long; lobes usually broadly ovate, 3–10 × 4–10 mm., rounded, entire, erect. Corolla pale yellow with the tube often pale green, creamy, or white, carnose, in the mature bud 29–33 mm. long, incl. the lobes (the lobes $\frac{2}{5}$–$\frac{3}{5}$ of the bud-length, being 13–19 × 10–13 mm., forming an almost conical head with a blunt apex), glabrous on both sides; tube as long as the calyx or slightly longer, almost cylindrical, 17–23 mm. long, twisted, widest around the ovary and there 5–8 mm. wide, contracted above, narrowest just below the insertion of the stamens and there 4–6 mm. wide, widened and thickened at the insertion of the stamens, inside with 5 small incrasations in the mouth (touching the connectives); lobes not twisted, broadly obcordate, 1·1–1·3 times as long as the tube, 1·2–1·5 times as wide as long, 19–30 × 28–43 mm., with basal portion about 3–5 × 3–5 mm., abruptly widened from there, at the apex emarginate, entire, spreading to recurved. Stamens exserted for 2–3 mm. (seemingly more in dry specimens), inserted 3–4 mm. below mouth of the corolla; anthers sessile, narrowly triangular, 6–7 × 3 mm., acuminate at the sterile apex (sterile part 2 mm.), glabrous. Pistil glabrous, approximately as long as the corolla tube, 17–23 mm. long; carpels 2, separate 2–3 × 2–3·5 × 1–1·7 mm., rounded, connected at the apex by the base of the style; style not split at the base, 12–19 × 1 mm., more or less gradually thickened at the apex; clavuncula capitate, 1–1·6 × 1–1·6 mm., with a ring 1·5–2 × 3·4 mm., being long-fimbriate at the base. Disk annular, lower or higher than the ovary, 2–4·5 mm. high, shallowly to deeply 5-lobed. Ovules about 80 in each carpel. Fruit of two free subglobose mericarps being slightly longer than wide and about as thick as wide, 4–10 cm. in diam., pale and dark green-spotted, also when mature, wall creamy inside and on section, 5(?)–15 mm. thick. Seeds with orange aril, dark brown, obliquely ovoid or ellipsoid, 8–9·5 × 6–6·5 × 4–5 mm., rough, densely papillose and shallowly grooved all over.

Zambia. N: Chinsali Distr., 40 km. S of Shiwa Ngandu, fl. 3.xii.1962, *Cottrell* 12 (SRGH). W: Mwinilunga Distr., N. of Dobeka Bridge, fl. 10.xi.1937, *Milne-Redhead* 3182 (B; BM; BR; EA; K; PRE; S). **Zimbabwe.** E: Chimanimani (Melsetter) Distr., end of Haroni Gorge, 11.vi.1971, *Müller & Gordon* 1851 (K; LISC; SRGH). **Malawi.** N: Nkhata Bay Distr., Nkuwazi F. R., 20 km. S. of Nkhata Bay, fr. 12.ix.1970, *Müller* 1612 (SRGH). S: Mulanje, 600 m., *Townsend* 65 (FHO). **Mozambique.** N: Serra de Ribaué, Nampula, fl. 2.ii.1937, *Torre* 1142 (COI; LISC). Z: Aringa, fl. 12.ii.1905, *Le Testu* 671 (BM; BR; G; MO; P; WAG). MS: Sofala Province (Beira), Chiniziua, near Gama, left bank Macalaua R., fl. 16.iv.1957, *Gomes e Sousa* 4364 (FHO; K). GI: Inhambane, Ave R., 20 km. S. of Inhambane, fl. i.1939, *Gomes e Sousa* 2197 (BR; COI; FI; LISC). M: Maputo, 15 km. Zitundo-Bela Vista Road, fl. & fr. 5.vii.1973, *Correia & Marques* 2911 (EA; K; LISC; LISU; LMU; M; SRGH; WAG).
    Widespread throughout tropical Africa inclusive of Madagascar; also in Swaziland and S. Africa (Natal). Moist places at margins of evergreen forest, in swampy forest and in riverine forest; 0–1600 m.

12. TABERNAEMONTANA L.

Tabernaemontana L., Sp. Pl. **1**: 210 (1753). — Stapf in F.T.A. **4**, 1: 150 (1902). Pichon in Not. Syst. **13**: 247 (1947). — in Mém. Mus. Nation. Hist. Nat., Paris Sér. 2, **27**: 225 (1949). — Leeuwenberg in Adansonia, Sér. 2, **16**: 390 (1976).
Pandaca Noronha ex Thouars., Gen. Nov. Madag.: 10 (1806). — Markgraf in Adansonia, Sér. 2, **10**: 29 (1970); in op. cit. **12**: 217 (1972); in Fl. Madag. Apocynac.: 180 (1976).
    Conopharyngia G. Don, Gen. Syst. **4**: 98 (1837). — Stapf, tom. cit. 139 (1902).
        Bonafousia A. DC., Prod. **8**: 359 (1844).
        Ochronerium Baill. in Bull. Soc. Linn. Paris **1**: 774 (1889).
Gabunia K. Schum. in Engler, Bot. Jahrb. **23**: 224 (1896). — Boiteau & Allorge in Bull. Mus. Nation. Hist. Nat., Paris, Sér. 4, **3**, Sect. B, Adansonia **2**: 216 (1981).
Ervatamia (A. DC.) Stapf, tom. cit.: 126 (1902). — Tsiang & Li in Fl. Rep. Pop. Sin. **63**: 98 (1977).
        Oistanthera Markgraf, tom. cit.: 550 (1972).
Hazunta Pichon in Not. Syst. **13**: 207 (1948). — Markgraf in Adansonia, Sér. 2, **10**: 27 (1970); op. cit. **12**: 222 (1972); in Fl. Madag., Apocynac.: 161 (1976).
    Muntafara Pichon, tom. cit. 209 (1948). — Markgraf, tom. cit.: 216 (1976).
        Pandacastrum Pichon, loc. cit. — Markgraf, tom. cit. 178 (1976).

Capuronetta Markgraf in Adansonia, Sér. 2, **12**: 61 (1972); tom. cit. 177 (1976).
  Sarcopharyngia (Stapf) Boiteau in Adansonia, Sér. 2, **16**: 272 (1976).
Camerunia (Pichon) Boiteau, tom. cit.: 274 (1976). — Boiteau & Allorge in Bull.
  Mus. Nation. Hist. Nat., Paris, Sér, 4, **3**, Sect. B. Adansonia **2**: 233 (1981).
  Leptopharyngia (Stapf) Boiteau, tom. cit. 276 (1976).
  Protogabunia Boiteau, loc. cit.

Shrubs or trees, repeatedly dichotomously branched and with 2 inflorescences in the forks when flowering (rarely one of the inflorescences not developing). Bark with abundant white latex. Branchlets terete or elliptic on section in African species, often sulcate and angular when dried. Leaves opposite, those of a pair equal or unequal, petiolate (at least in species of the F.Z. area); petioles of a pair connate into a conspicuous ocrea, with colleters in the axils; lamina broadly to narrowly elliptic. Inflorescence distinctly pedunculate, corymbose, lax to congested. Bracts deciduous, but at least in species of the F.Z. area still present at anthesis. Flowers actinomorphic, except for the subequal sepals, fragrant. Sepals green or of different colour, almost free. Corolla white, pale yellow or mauve (Madagascan spp.), with an often green or greenish tube and a pale yellow throat, thick and carnose or thin; tube twisted, in species of the F.Z. area at least twice as long as the calyx; lobes in bud overlapping to the left and folded inwards, twisted, usually obliquely elliptic, usually curved to the right, spreading and often recurved later. Stamens included or less often exserted; filaments reduced to ridges or nearly so; anthers narrowly triangular, acuminate at the sterile apex, sagittate at the base, introrse, dehiscent throughout by a longitudinal slit. Ovary composed of two carpels, these barely to distinctly connate at the base and connected at the apex by the style; style often split at the base; disk absent (at least in species of the F.Z. area); clavuncula not coherent with the anthers and therefore the pistil still complete immediately after the corolla is shed. Fruit of two free or basally united subglobose, subellipsoid or pod-like carnose carpels (sometimes one remaining smaller or not developing); carpels opening along an adaxial line of dehiscence or not. Seeds several or many, enveloped by a thin variously coloured pulpy aril, nearly obliquely ellipsoid, with a deep groove to halfway its width at the hilar side; endosperm ruminate.

A circumtropical genus comprising about 120 species.

1. Flowers small; sepals 1·2–2·5 mm. long; corolla tube 5–7 mm. long; carpels obliquely ovoid
   or obliquely ellipsoid, with 3 ridges and pale brown warts   -   -   -   - 1. *elegans*
 - Flowers large; sepals 3·5–7 mm. long; corolla tube at least 10 mm. long; carpels subglobose
   or obliquely ellipsoid, smooth or dotted, without warts -   -   -   -   -   - 2
2. Corolla tube 10–27 mm. long, slightly twisted at the base, glabrous inside above the
   insertion of the stamens at 6–8 mm. from the corolla base; lobes in the mature bud forming a
   head not wider than the widest part of the tube; leaves narrowly elliptic, 4–27 × 1·5–10 cm.
   often undulate or sinuate; carpels obliquely ellipsoid, 6–7(10) × 4·5–5(7) × 4–4·5(6) cm.,
   smooth; aril orange -   -   -   -   -   -   -   -   -   - 4. *ventricosa*
 - Corolla tube 18–42 mm. long — if clearly twisted — anthers inserted 11–15 mm. above the
   base, sparsely to densely pubescent or pilose above the insertion of the stamens; lobes in the
   mature bud forming a head much wider than the widest part of the tube; leaves broadly or
   narrowly elliptic, 10–50 × 3–26 cm.; carpels subglobose, about 7–20 cm. in diam.; aril white
   (colour not yet known for *T. stapfiana*) -   -   -   -   -   -   -   - 3
3. Fresh corolla tube 5-angular, not or barely twisted, inside densely or (less often) sparsely
   pilose from above the insertion of the stamens to the basally pubescent lobes, with appressed
   pubescence directed downwards below the anthers; at the base of the inside of the sepals 1–3
   rows of small colleters, about 0·5 × 0·2 mm.; seeds 11–14 mm. long; secondary veins of leaf
   lamina forming an angle of 60–80° with the midrib and more gradually upcurved than in the
   following species-   -   -   -   -   -   -   -   -   - 2. *pachysiphon*
 - Fresh corolla tube not angular, a whole turn twisted over its entire length, inside sparsely
   pubescent in a belt stretching from below or above the insertion of the stamens to the mouth;
   at the base of the inside of the sepals a single row of 6–10 large colleters (1)1·5 × 0·3–0·5 mm.;
   seeds 15–21 mm. long; secondary veins of leaf lamina forming an angle of 70–90° with the
   midrib and rather straight for about two thirds of their length -   -   - 3. *stapfiana*

1. **Tabernaemontana elegans** Stapf in Kew Bull. **1894**: 24 (1894). — Codd in Fl. Southern
   Afr. **26**: 270, fig. 39 (1963). — Leeuwenberg in Adansonia, Sér. 2, **16**: 387, pl. 2.4 (1976).
   Type: Mozambique, Delagoa Bay, *Monteiro* 55 (K, lectotype; P, isotype).
     Conopharyngia elegans (Stapf) Stapf in F.T.A. **4**, 1: 149 (1902). Type as above.
     Leptopharyngia elegans (Stapf) Boiteau in Adansonia, Sér. 2, **16**: 276 (1976). Type as
   above.

Small tree or shrub, 1·5–12 m. high; trunk terete, 5–30 cm. in diam.; bark pale brown, corky, deeply fissured. Branches with a pale brown corky longitudinally fissured bark and transverse ridges caused by leaf scars, with scattered lenticells of the same colour; branchlets glabrous. Leaves petiolate; petiole glabrous, 7–30 mm. long; lamina coriaceous (also when fresh) variable in size, elliptic to narrowly elliptic, 2–4 times as long as wide, (4)5·5–23 × (1)2–8 cm., acuminate with an often blunt point or acute or obtuse, with the base cuneate or decurrent into the petiole, entire, glabrous on both surfaces or occasionally pubescent beneath; secondary veins 12–23 on either side of the midrib, straight except for the upcurved apex, fish-bone-like, forming an angle of 70–90° with the midrib, anastomosing towards the apex; tertiary venation reticulate, especially conspicuous in old leaves. Inflorescences short- or long-pedunculate, 5–20 × 5–15 cm., usually many-flowered, lax. Peduncle glabrous or occasionally with scattered short hairs, 1–8·5 cm. long, rather slender; pedicels glabrous or occasionally with scattered short hairs, 2–6 mm. long. Bracts scale-like, about as long as the sepals, soon deciduous, with a single row of colleters in the axils. Flowers sweet-scented. Sepals pale green, almost free, subcircular or broadly ovate, 0·8–1 times as long as wide, 1·2–2·5 × 1·2–2·5 mm., glabrous or occasionally pubescent outside, inside at the base with a single row of colleters (this may be interrupted and restricted to groups of 3–4, sometimes basally united ones), rounded, erect. Corolla white, creamy, or pale yellow, in the mature bud 8·5–15 mm. long incl. the lobes (the lobes ⅓–c. ½ of the length of the bud and 3·5–8 × 3·5–6 mm., forming a broadly ovoid head with a blunt apex), glabrous or occasionally with some small hairs outside, inside with a densely pubescent belt from the insertion of the stamens to the mouth; tube 2·5–4 times as long as the calyx, 5–7 mm. long, almost cylindrical and 1·8–2·4 mm. wide, not twisted, slightly contracted at both the base and just below the insertion of the stamens, widened at the throat; lobes slightly falcate, 1·3–2·5 times as long as the tube, 1·5–3·3 times as long as wide, 8–15 × 3–7 mm., rounded, subauriculate at the left side of the base, entire. Stamens included for 1–2 mm., inserted 2–2·7 mm. above the corolla-base; anthers 2–2·5 × 0·6–0·7 mm., glabrous. Pistil glabrous, 3·5–4·2 mm. long; ovary subglobose to almost cylindrical, laterally compressed, 1·2–1·6 × 1·2–1·4 × 1–1·2 mm.; style 1–1·2 × 0·2 mm., widened at the apex; clavuncula composed of two rings: a narrow basal one, 0·5–0·8 × 0·1 mm. and a wider apical one, 0·3–0·4 × 0·5–0·6 mm. gradually narrowing into the slender (0·5)1 mm. long erect stigma. Ovules c. 35–60 in each carpel. Mericarps separate, glaucous or green, with conspicuous pale brown warts, obliquely ovoid or ellipsoid, 5–8 × 4–6·5 × 4–5 cm., apiculate, with 3 ridges, two lateral and one abaxial (these more pronounced in dried fruits), 2-valved, several-seeded; wall 5–15 mm. thick; aril orange. Seeds dark brown, dull, 14–15 × 7–9 × 6–7 mm., with reticulate grooves making a brain-like surface, minutely tuberculate.

Malawi. S: Nsanje (Port Herald), Thangadzi R., 80 m., fr. 25.iii.1960, *Phipps* 2723 (MO; SRGH). **Mozambique.** N: Niassa, Nampula, Monapo and Mutivaze Rs., fl. 25.i.1936 *Torre* 673 (COI; LISC). Z: 50 km. N.E. of Mopeia Velha on Road to Quelimane, 60 m., fl. 7.xii.1971, *Müller & Pope* 1951 (LISC; LMA; SRGH). MS: Cheringoma, Inhaminga, fl. 10.x.1944, *Simão* 160 (LISC; LMA). GI: Gaza, Barra do Limpopo, 3 km. de Gumbe, fl. 26.v.1965, *Pereira et al.* 454 (LMU; WAG). M: between Costa do Sol and Marracuene, fl. 12.xi.1960, *Balsinhas* 240 (COI; LISC; LMA).
Also known from Somalia, E. and S. Africa. Woodland, usually with *Brachystegia*, frequent near the coast and there on dunes, less so inland and then in riverine forests. More or less fire-resistant by its corky bark; 0–700 m.

2. **Tabernaemontana pachysiphon** Stapf in Kew Bull. **1894**: 22 (1894). — H. Huber in F.W.T.A., ed. 2, 2: 66 (1963). TAB. **101**. Type from Nigeria.
   Tabernaemontana angolensis Stapf, tom. cit. 23 (1894). Type from Angola
   Tabernaemontana holstii K. Schum. in Engler, Pflanzenw. Ost-Afr. **C**: 317 (1895). Type from Tanzania.
   Voacanga dichotoma K. Schum., loc. cit. Type from Tanzania.
   Conopharyngia cumminsii Stapf in F.T.A. **4**, 1: 145 (1902). Type from Ghana.
   Conopharyngia pachysiphon (Stapf) Stapf, tom. cit. 146 (1902). Type as for *T. pachysiphon*.
   Conopharyngia holstii (K. Schum.) Stapf, loc. cit. Type as for *T. holstii*.
   Conopharyngia angolensis (Stapf) Stapf, loc. cit. Type as for *T. angolensis*.
   Tabernaemontana pachysiphon var. *cumminsii* (Stapf) H. Huber in Kew Bull. **15**: 438 (1962). Type as for *C. cumminsii*.

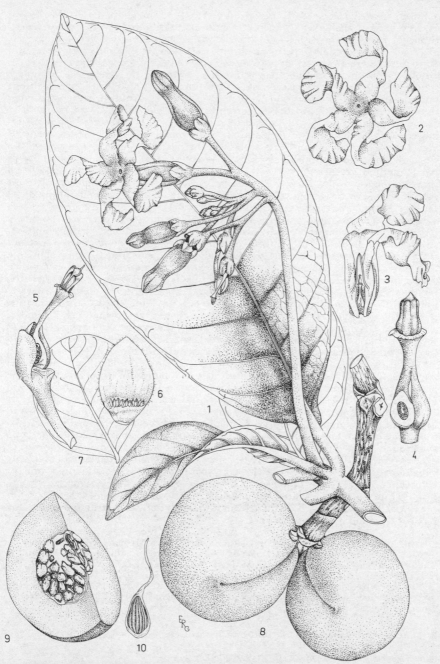

Tab. 101. TABERNAEMONTANA PACHYSIPHON. 1, flowering branch (× ½); 2, flower (× ½); 3, opened corolla with pistil (× ½); 4, pistil with longitudinal section of one carpel (× 3), 1–4 from *Leeuwenberg* 11242; 5, calyx with pistil (× 1½); 6, sepal inside, with colleters (× 3), 5–6 from *Leeuwenberg* 11933; 7, leaf (× ¼), from *Okafor & Latilo* FHI 57181; 8, fruiting branch (× ¼), from *Leeuwenberg* 11903; 9, section of fruit (× ½), from *Lap* 239; 10, seed (× 1), from *Leeuwenberg* 11044.

Shrub or small tree 2–18 m. high, trunk terete, 4–40 cm. in diam. Bark pale brown or grey-brown, shallowly to deeply longitudinally fissured, with large lenticels. Branches pale brown, with large lenticels, with conspicuous leaf scars; branchlets glabrous. Leaves petiolate; petiole glabrous, 6–19 mm. long; lamina coriaceous (also when fresh), variable in size, broadly to narrowly elliptic, 1·3–3 times as long as wide, 10–50 × 5–26 cm., with the apex acuminate, acute, or in some leaves rounded, rounded or cuneate at the base, entire, with a more or less revolute margin, glabrous on both surfaces; secondary veins 7–16 on each side of the midrib, rather straight at the base and gradually upcurved, forming an angle of 60–80° with the midrib. Inflorescences mostly long-pedunculate, suberect, 8–26 × 5–15 cm., few- to many-flowered, fairly lax. Peduncle glabrous, 3–14 cm. long, fairly robust; pedicels glabrous, 8–22 mm. long. Bracts scale-like, much smaller than the sepals, with a single row of colleters in the axils. Flowers sweet-scented, open in day time. Sepals pale to medium green, almost free, thick, carnose, erect, subcircular or ovate, 0·8–1·4 times as long as wide, 4–7 × 4–7 mm., imbricate in bud, entire, ciliate, glabrous outside, inside with 1–3 dense rows of small colleters closely together in the middle of the base. Corolla white, sometimes pale yellow, with a pale green tube and often with a pale yellow throat, thickly carnose, in the mature bud with a ventricose tube and a comparatively small blunt broadly ovoid head formed by the lobes (being about ⅕ of the whole length of the bud), outside puberulous only on the bud covered part and there mostly only at the base, rarely entirely glabrous, inside from 0–3 mm. above the base and reaching to 0–2 mm. below the insertion of the stamens with an incomplete (interrupted at the centre of the anthers) belt of soft short downward pointing hairs, between the insertion of the stamens and the mouth densely to (sometimes) sparsely pilose, above the mouth pubescent up to 0·3–0·6(1) times the length of the lobes; tube 3–5(6) times as long as the calyx, 18–35(42) mm. long, almost cylindrical, 5-angular, at the base 4–8 mm., ventricose at the insertion of the stamens or barely below and there 8–13 mm., above 5–10 mm. wide, not or (especially at the base) slightly twisted (up to ¼ turn); lobes usually more or less falcate, 0·7–1·6 times as long as the tube, 1·5–4 times as long as wide, 14–50 × 6–18(27) mm., rounded, not auriculate, undulate. Stamens barely to up to 12 mm. included, inserted ¼–½ the length from the base of the corolla tube, being 8–14 mm.; anthers 9–13 × 4–5 mm., glabrous. Pistil glabrous, 15–20 mm. long; ovary almost cylindrical, 3–5 × 3–4 × 3–4 mm.; style 7–10 × 0·8–1·4 mm., slightly widened at the apex; clavuncula almost cylindrical, 3–4·5 mm. long, laterally obscurely 5-angular and at the base widened into a 2–3·5 mm. wide ring, in the middle 1·2–1·5 mm. in diam., at the apex widened and with 5 subcircular lobes altogether 1·5–2 mm. wide; stigma 0·3–1 mm. long, obtuse or acute. Ovules c. 100 in each carpel. Mericarps pale glaucous, often dotted, obliquely subglobose, 7 × 6 × 6–15 × 13 × 14 cm., rounded, with an indented line of dehiscence, but probably indehiscent, several- to many-seeded; wall on section and inside white, 2·4 cm. thick; aril white. Seeds dark brown, dull, 11–14 × 5–6·5 × 4·5–7 mm., with longitudinal grooves, with a minute honeycomb-like structure covered by pale brown minute deciduous warts.

Zambia. N: Mansa (Fort Rosebery) Distr., near Samfya Mission, Lake Bangweulu, fl. 30.viii.1952, *White* 3181 (BR; FHO). W: Mwinilunga Distr., near Kalene Hill Mission, fl. 27.ix.1952, *White* 3390 (FHO; WAG). **Malawi.** N: Nkhata Bay Distr., Chombe Tea Estate 600 m., imm. fr. 22.ii.1976, *Pawek* 10869 (BR; K; PRE; SRGH; UC; WAG).

Also known from most other countries of tropical Africa, from Ghana to Sudan in the north, to Angola in the southwest and Tanzania in the east. Light forest understorey and riverine forest; 0–1500 m.

3. **Tabernaemontana stapfiana** Britten in Trans. Linn. Soc., Ser. 2, **4**: 25 (1894). Type: Malawi: Mulanje Mt., *Whyte* 87 (BM, holotype).
   Conopharyngia stapfiana (Britten) Stapf in F.T.A. **4**, 1: 147 (1902). Type as above.
   Conopharyngia johnstonii Stapf, loc. cit. Type from Uganda.
   Conopharyngia bequaerti De Wild., Pl. Bequaert. **1**: 397 (1922). Type from Zaire.
   Conopharyngia johnstonii var. *grandiflora* Markgraf in Notizbl. Bot. Gart. Berlin **8**: 496 (1923). Type from Kenya.
   Tabernaemontana johnstonii (Stapf) Pichon in Not. Syst. **13**: 251 (1948). Type as for *Conopharyngia johnstonii*.
   Sarcopharyngia stapfiana (Britten) Boiteau, Bull. Mus. Nation. Hist. Nat. Paris, Sér. 4, **3**: 233 (1981), partly, excl. of syn. *Gabunia odoratissima* Stapf. Type as for *Tabernaemontana stapfiana*.

Tree 5–25(35) m. high; trunk terete, 25–90 cm. in diam.; bark pale to dark grey-brown, rough (?), 1 cm. thick, corky. Branches shallowly longitudinally fissured when dried, with large lenticels of the same colour, with conspicuous leaf scars; branchlets glabrous. Leaves petiolate; petiole glabrous, 5–30 mm. long; lamina coriaceous (also when fresh), variable in size, mostly narrowly elliptic, 2–4·5 times as long as wide, 12–40 × 3–14 cm., acuminate, apiculate, or rounded, at the base cuneate or decurrent into the petiole, entire or sometimes sinuate, often with slightly revolute margins, glabrous on both surfaces; secondary veins 12–24 on each side of the midrib, fairly straight at the base, upcurved, forming an angle of 70–90° with the midrib. Inflorescences often long-pedunculate, 10–28 × 7–15 cm., few- to many-flowered, fairly lax. Peduncle glabrous, 3–15 cm. long, fairly robust; pedicels glabrous, 5–30 mm. long. Bracts scale-like, much smaller than the sepals, with a row of colleters in the axils. Flowers sweet-scented, open in day time. Sepals medium green, almost free, thick, carnose, erect, subcircular to oblong, 1–1·5 times as long as wide, 5–7 × 4–7 mm., often ciliate, outside glabrous, inside closely together in the middle of the base with one row of 6–10 large colleters (1)1·5 × 0·3–0·5 mm. Corolla white, with a pale green tube and often with a pale yellow throat, thickly carnose, in the mature bud with a ventricose tube and a comparatively large ovoid head formed by the lobes (being conspicuously wider than the tube and c. ⅓–⅔ of the length of the bud, with a subacute point), outside glabrous or with a small patch of pubescence at the left side of the base of the lobes (and occasionally with 5 stripes 7 mm. long of pubescence on the tube below), inside pilose in a belt stretching from 6 mm. below to 2 mm. above the insertion of the stamens and ending barely below the mouth or reaching the base of the lobes; tube 3–4·5 times as long as the calyx, 21–42 mm. long, almost cylindrical, not angular, at the base 2·5–5 mm., ventricose at the insertion of the stamens or barely below and there 5·5–9 mm., above 3·5–6 mm. wide, 0–¼ turn twisted over the entire length of the tube; lobes not always falcate, 0·7–1·8 times as long as the tube, 1·5–2·3 times as long as wide, 17–50 × 8–24 mm., rounded, not or obscurely auriculate at the left side of the base, undulate, spreading and recurved later. Stamens 2–15 mm. included, inserted at ⅓–½ of the length from the base of the corolla tube (being at 11–15 mm. from the base); anthers 9–12·5(17) × 3–4 mm., glabrous. Pistil glabrous, 13–17 mm. long; ovary almost cylindrical, 4–6 × 2–4 × 2–3 mm., tapering into the style; style 5–10 × 0·8–1 mm., slightly widened at the apex; clavuncula almost cylindrical, 2–4·5 mm. long, at the sides obscurely 5-angular, and at the base widened into a 1·5–3 mm. wide ring, in the middle 1–1·2 mm. in diam., at the apex widened into 5 subcircular lobes altogether c. 1·2–1·7 mm. wide; stigma bilobed, 0·2–1 mm. long obtuse or acute. Ovules 100–200 in each carpel. Mericarps separate, dark green, with pale green, white, or yellow dots, very big, subglobose, 10–20 × 8–20 × 8–20 cm., rounded, with an indented line of dehiscence, but probably indehiscent, several- to many-seeded; wall on section and inside white(?), 2·5–6 cm. thick; aril white(?). Seeds dark brown, 15–21 × 9–12 × 7–10 mm., with longitudinal and transverse or oblique grooves, with a minute honeycomb-like structure.

Zimbabwe. E: Himalayas, Banti North, 2100 m., fr. 5.iii.1955, *Wild* 4508 (LISC; MO; S; SRGH). **Malawi.** N: Chitipa Distr., Misuku Hills, Mughesse Forest, 1750 m., fl. 28.xii.1972, *Pawek* 6185 (K; UC; WAG). South Vipya, Mthungwa Forest, fl. 7.i.1967, *Hilliard & Burtt* 4338 (E). S: Mulanje Distr., Mulanje Mt., Big Ruo Gorge, 1500 m., fr. 7.vii.1958, *Chapman* 602 (FHO; MO; SRGH). **Mozambique.** N: Cabo Delgado; Macondes, 5 km. Chomba-Negomano Road, 800 m., imm. fr. 13.iv.1964, *Torre & Paiva* 11986 (LISC). Z: Gúruè, near source of Malema R., 1750 m., 4.i.1968, *Torre & Correia* 16897 (LISC). MS; Báruè, Choa Mts., 17 km. from Catandica (Vila Gouveia), 1500 m., fl. 13.xii.1965, *Torre & Correia* 13602 (LISC).

Also known from East Africa. Montane forest; 700–2500 m.

4. **Tabernaemontana ventricosa** Hochst. ex A. DC., Prod. **8**: 366 (1844). — Codd in Fl. Southern Afr. **26**: 269, fig. 39, 2 (1963). Type from S. Africa.

Tabernaemontana usambarensis K. Schum. ex Engl. in Abh. Preuss. Akad. Wiss. **1894**: 36 (1894). — K. Schum. in Engl. & Prantl, Nat. Pflanzenfam. **4**, 2: 148 (1895). Type from Tanzania.

Conopharyngia usambarensis (K. Schum.) Stapf in F.T.A. **4**, 1: 148 (1902). Type as for *T. usambarensis*.

Conopharyngia ventricosa (Hochst. ex A. DC.) Stapf, tom. cit.: 149 (1902). Type as for *T. ventricosa*.

Conopharyngia rutshurensis De Wild., Pl. Bequaert. **1**: 399 (1922). Type from Zaire.
Sarcopharyngia ventricosa (A. DC.) Boiteau in Adansonia, Sér. 2, **16**: 272 (1976). Type as for *T. ventricosa*.

Shrub or small tree 3–15 m. high. Trunk terete, 5–30 cm. in diam.; bark pale brown, shallowly to deeply longitudinally fissured, often slightly corky, with large lenticels. Branches pale brown, shallowly longitudinally fissured and with transverse ridges caused by leaf scars when dried, with large lenticels of the same colour; branchlets glabrous. Leaves petiolate; petiole glabrous, 3–15 mm. long, with a single row of colleters in the axils; lamina coriaceous (also when fresh), variable in size, narrowly elliptic, 2·5–4 times as long as wide, 4–27 × 1·5–10 cm., acuminate with a blunt point, acute, or obtuse, cuneate or rounded at the base, entire, sinuate, or undulate, glabrous on both surfaces; secondary veins 7–23 on each side of the midrib, straight or slightly curved, forming an angle of 60–80° with the midrib anastomosing at the apex; tertiary venation reticulate, inconspicuous. Inflorescences usually long-pedunculate, 5–23 × 3–10 cm., usually many-flowered, more or less congested. Peduncle glabrous, 2–15 cm. long, fairly robust; pedicels glabrous, 3–10 mm. long. Bracts scale-like, almost as long as the sepals, with a single row of colleters in the axils. Flowers sweet-scented. Sepals pale green, almost free, thick, carnose, erect, subcircular or broadly ovate, 1–1·5(2) times as long as wide, 3·5–6 × 2–5 mm., ciliate, outside glabrous, inside with a single row of colleters at the base sometimes being interrupted and restricted to groups of 4–6 in front of each sepal. Corolla white, often with a pale yellow throat and a greenish tube or less often entirely pale yellow, in the mature bud 12–35 mm. long incl. the lobes (the lobes $\frac{1}{7}$–$\frac{1}{2}$ of the length of the bud, 4–10 × 4–8 mm., forming a broadly ovoid head with a blunt apex), outside glabrous, inside from 1·5–2·5 mm. above the base with an incomplete (interrupted below the centre of the anthers) 2–4 mm. wide belt of soft appressed hairs; tube 2·1–5·4 times as long as the calyx, 10–27 mm. long, almost cylindrical to slenderly bottle-shaped, variable, at the base 2·5–4 mm., ventricose at the insertion of the stamens or just below and there 4–6 mm. and at the throat or just below 2·5–4 mm. wide, slightly twisted only just above the base; lobes usually clearly falcate, 0·7–2 times as long as the tube, 14–32 × 5·5–16 mm., rounded, auriculate at the left side of the base, undulate. Stamens barely exserted to for 14 mm. included, inserted 6–8 mm. above the corolla base; anthers 5·5–9·5 × 2–2·7 mm., glabrous. Pistil glabrous, 7–10 mm. long; ovary almost cylindrical to broadly ovoid, often laterally compressed, 2–4·5 × 1·5–3 × 1·4–2·5 mm.; style 2·5–5·3 × 0·2–0·3 mm., slightly widened at the apex; clavuncula almost cylindrical, 2·5–3 × 1–1·2 mm., widened at the base into a thin 1·2–1·7 mm. wide ring, apically gradually narrowing into 5 subglobose lateral 0·5–0·7 mm. long lobes; stigma 0·5–0·7 mm. long. Ovules c. 70–100 in each carpel. Mericarps separate, dark green, obliquely ellipsoid, 6–7 × 4·5–5 × 4–4·5 cm. (or perhaps larger), rounded, with 2 faint lateral ridges (usually visible in the often wrinkled dried fruit), smooth, dehiscent, several-seeded; wall 7–13 mm. thick, white inside; aril orange. Seeds dark brown, dull, 11–13 × 4–7 × 3–5 mm., with longitudinal grooves, minutely tuberculate.

Zimbabwe. E: Mutare (Umtali) Distr., Vumba Mts., Witchwood Estate, 1400 m., fl. 12.xii.1955, *Drummond* 5082 (COI; LISC; LMA; SRGH). **Malawi. S:** Mulanje Mt., fr. 14.vii.1958, *Chapman* 607 (FHO). **Mozambique. N:** Malema, km. 40 from Entre Rios on the road to Ribáuè, Murripa Mts., 1100 m., immat. fr. 15.xii.1967, *Torre & Correia* 16518 (LISC). **MS:** Mt. Espungabera (Spungabera), fl. 21.xi.1960, *Leach & Chase* 10501 (COI; LISC; LMA; MO; SRGH).

Also known from Central and East Africa. Light or secondary forest, woodland and there often in thickets or riverine forest; 0–1850 m.

13. CARVALHOA K. Schum.

Carvalhoa K. Schum. in Engler & Prantl, Nat. Pflanzenfam. **4**, 2: 189 (1895).

Shrub or small tree, repeatedly dichotomously branched and with 2 inflorescences in the forks when flowering, with white latex. Branchlets terete. Leaves opposite, those of a pair equal or unequal, shortly petiolate or sometimes sessile; petioles or leaf-bases of a pair united at the base and forming a very short ocrea, with a single row of colleters in the axils; lamina elliptic, narrowly elliptic, or narrowly obovate, cuneate, rounded or subcordate at the base. Inflorescence pedunculate, corymbose, lax. Bracts caducous before the flowers open. Flowers actinomorphic except for the slightly curved corolla tube. Calyx pale green, persistent, even under the fruit. Corolla white, creamy, or pale yellow, with many red longitudinal lines at the base of the lobes and at the apex of the tube; tube longer than the calyx; lobes spreading(?) or suberect in bud overlapping to the left. Stamens included; anthers sessile, narrowly triangular, acuminate at the sterile apex, sagittate at the base, hirto-pubescent outside on the connective and inside between the tails, introrse. Pistil glabrous; ovary broadly ovoid, composed of 2 free carpels, surrounded by an entire disk, which is united with the distal sides of the ovary; style only slightly widened at the apex; clavuncula composed of an entire, thin, slightly recurved ring and a subglobose head, topped by 2 parallel linear erect stigma-lobes. Style and stigma remaining on the ovary when the corolla is shed. Fruit composed of 2 free carpels which usually both develop and which dehisce adaxially; wall soft, orange inside. Seeds surrounded by a darker orange pulpy aril, few, rather large, obliquely ellipsoid or ovoid; at the hilar side with one deep groove to half their width, less deeply grooved at the other sides, minutely pustulate; endosperm copious, starchy, creamy, ruminate, surrounding the spathulate embryo.

A monotypic African genus.

Carvalhoa campanulata K. Schum. in Engler & Prantl, Nat. Pflanzenfam. **4**, 2: 189 (1895). TAB. **102**. Type: Mozambique, Mussovil, Cabraceira, 1884–1885 *Carvalho* (B†; COI, lectotype (by Leeuwenburg); K, P, Z, isotypes).
  C. macrophylla K. Schum. in Engler, Bot. Jahrb. **30**: 381 (1901). Type from Tanzania.
  C. petiolata K. Schum. in Engler, Bot. Jahrb. **33**: 317 (1903). Type from Tanzania.

Shrub or small tree 1–5 m. high. Trunk 20 cm. in diam. (teste Robson 1657). Branches pale grey-brown, lenticellate, with shallowly longitudinally fissured bark; branchlets glabrous or pubescent, sulcate when dry, lenticellate. Leaves petiolate; petioles glabrous or less often pubescent, 1–7 mm. long, lamina membraneous when dry, variable in shape and size 2–4(5) times as long as wide, 4–26 × 1·5–12 cm., acuminate and sometimes with a blunt tip at the apex, symmetric or asymmetric and cuneate, rounded, or subcordate at the base, if asymmetric often cuneate on one and rounded or subcordate on the other side, entire, glabrous, with a few hairs, or less often pubescent on both surfaces; secondary veins 8–15 on each side, arcuate and anastomosing with each other and with the reticulate tertiary venation. Inflorescences pendulous, 4–17 × 3–10 cm., 2–4 times branched, partly more or less dichasial. Peduncle very thin, 1·5–7 cm. long, glabrous or less often sparsely pubescent as both branches and pedicels; pedicels thin, 5–20 mm. long, thickened at the apex. Bracts minute, sepal-like, without or with a few axillary colleters leaving large leaf-scars. Sepals connate at the base, 1–1·7 times as long as wide, 1·5–3·5 × 1·3–2·4 mm. triangular or ovate, acuminate or acute, imbricate in bud, erect, glabrous, puberulous or pubescent outside, with 0–3 colleters above the base inside near both edges or occasionally with 7 colleters in a single row at the same level (the number may vary within a single flower from 0–3), ciliate or ciliolate. Corolla in the mature bud about 10 mm. long, of which the lobes are about ¼ and do not shape a head, glabrous or puberulous and then with a glabrous base outside, inside hirto-pubescent between and often just below the anthers and above sometimes puberulous; tube 2·5–7 times as long as the sepals, 8–10 m. long, slightly constricted at 4–4·5 mm. and there 2–3 mm. wide and above gradually widened towards the throat and there 6–9 mm. wide; lobes subcircular, 3–6 × 3·5–6 mm., rounded, entire. Stamens inserted 2·8–3·5 mm. above the corolla-base; anthers sessile (even lower half of connective connate with the corolla tube by a cushion), 3·4–3·7 × 1·2–1·5 mm., with a sterile apex. Pistil 4·5–5 mm. long; ovary 2–2·5 × 1·5–2 × 0·8–1·5 mm.; disk entire, 0·3–1 mm. high; style split at the base often for even more than half its length, cylindrical, 1·2–1·5 mm. long; clavuncula ring 1–1·2 mm. in diam., and head 0·8 mm. in diam., stigma-lobes 0·2–0·4 mm. long and about 0·02–0·04 mm. thick. Ovules c. 30 in 4 rows on one oblong placenta in each cell. Follicles yellow or pale orange or less often green outside, pod-like, recurved or straight, 2–7 times as long as wide, 3–6 × 0·8–1 cm., acuminate at the apex, rounded or cuneate at the base, smooth when fresh, indented around the seeds only when dry, dehiscent throughout by an

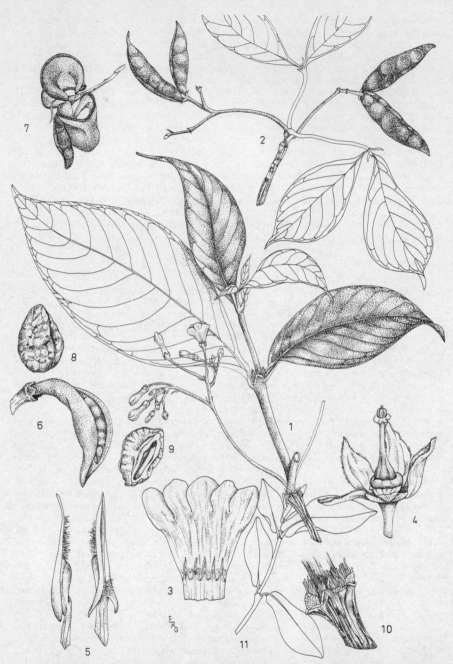

Tab. 102. CARVALHOA CAMPANULATA. 1, flowering branch ($\times \frac{1}{2}$), from *Goetze* 1343; 2, fruiting branch ($\times \frac{1}{2}$), from *Baag* et al. 93; 3, opened corolla ($\times 3$); 4, calyx with pistil ($\times 5$); 5, stamens ($\times 8$), 3–5 from *Schlieben* 5454a; 6, dehiscing fruit ($\times 1$), from *Drummond &* *Hensley* 1703; 7, open fruit, bicarpellate ($\times \frac{1}{2}$); 8, seed ($\times 3\frac{1}{2}$); 9, longitudinal section of seed ($\times 3\frac{1}{2}$); 10, node with colleters ($\times 3$), 7–10 from *Milne-Redhead &* *Taylor* 935; 11, branchlet ($\times \frac{1}{4}$), from *Mendonça* 1136.

adaxial longitudinal slit and then becoming flat and up to 7·5 × 2·5 cm. Seeds in 2 irregular rows, pale brown, laterally compressed, 5·5–6 × 4–4·5 × 2·5–3 mm., flat at one side, minutely pustulate; embryo straight, white.

Malawi. N: Walindi Forest, Misuku Hills, 2000 m., fl. 12.xi.1958, *Robson & Fanshawe* 577 (BM; BR; LISC; SRGH). C: Ntchisi Mt., 1450 m., imm. fr. 19.ii.1959, *Robson & Steele* 1657 (BM; LISC; SRGH). S: Mulanje Distr., Ruo Gorge, about 900 m., fl. 2.ix.1970, *Müller* 1486 (PRE; SRGH). **Mozambique.** N: near Mambiti, Mueda-Negomano road, fr. 2.iv.1960, *Gomes e Sousa* 4561 (COI; PRE; SRGH). Z: Milange, Serra de Tumbine, fl. 12.ix.1942, *Mendonça* 1401 (LISC).

Also known from Kenya and Tanzania. Montane rain forest or secondary forest; (300)800–1900 m.

14. SCHIZOZYGIA Baill.

Schizozygia Baill. in Bull. Soc. Linn. Paris **1**: 752 (1888); Hist. Pl. **10**: 202 (1889). — K. Schum. in Engler & Prantl, Nat. Pflanzenfam. **4**, 2: 109 (1895). — Barink in Meded. Landb. Wag. **83**–7: 47, fig. 7 (1984).

Shrub or small tree, repeatedly dichotomously branched and with 2 inflorescences in each fork, with white latex; branchlets terete. Leaves opposite, those of a pair equal, petiolate, with one or two rows of colleters in the axils; lamina obovate, cuneate at the base. Inflorescences corymbose, congested. Flowers actinomorphic except for the sometimes slightly unequal sepals. Calyx green. Corolla with tube yellow and limb creamy to yellow; tube slightly shorter to much longer than the calyx; lobes in the bud overlapping to the right, spreading. Stamens included or nearly so; anthers sessile, narrowly triangular, acute at the apex, sagittate at the base, glabrous or sometimes papillose, introrse, with upper two-thirds fertile. Pistil glabrous or nearly so; ovary subglobose, composed of 2 free carpels; disk adnate to the ovary for ± ⁷⁄₁₀ of its height; clavuncula cylindrical; stigma minute, 2-lobed. Fruit composed of 2 almost free adaxially dehiscent carpels, wall thinly coriaceous, irregularly striate, grooved when dry. Seeds few, surrounded by a thin red or orange pulpy aril, obliquely ellipsoid, with a deep groove to ½ their width at the hilar side and shallowly grooved at the other sides, minutely rugose; endosperm copious, starchy, white, ruminate, surrounding the spathulate embryo.

A monotypic African genus.

Schizozygia coffaeoides Baill. in Bull. Soc. Linn. Paris, **1**: 752 (1888); Hist. Pl. **10**: 202 (1889). — K. Schum. in Engler & Prantl, Nat. Pflanzenfam. **4**, 2: 109 (1895). — Renner in Lloydia **27**: 406–415 (1964). — Barink in Meded. Landb. Wag. **83**–7: 49, fig. j, map 7 (1984). TAB. **103**. Type from Tanzania.

Shrub or tree 1–4(8) m. high. Bark rough, brown, with pale lenticels, inside red; wood soft, pale yellow. Branches yellow to dark brown, lenticellate; branchlets often sulcate when dry, glabrous. Leaves petiolate; petiole 0·5–9 mm. long, glabrous; lamina 1·1–4·4 times as long as wide, 2·4–25 × 1·1–11 cm., acuminate, subcoriaceous when dry, glabrous on both sides; secondary veins conspicuous. Inflorescences 7–15 × 12–18 mm. Bracts narrowly oblong, 3–5 mm. long, acute, glabrous on both sides. Peduncle very short, up to 3 mm. long, glabrous; pedicels 2–3 mm. long, glabrous. Flowers fragrant. Sepals free, subequal, entire, imbricate, elliptic, acute or acuminate, 1·2–3·3 times as long as wide, 3·1–6 × 1·5–3·9 mm., outside glabrous, inside with 5–10 colleters, 0–2 near the base at the edges of each sepal. Corolla 1·3–4·7 times as long as the calyx, 6·5–9·5 mm. long, hypocrateriform; tube cylindrical or urceolate, 0·8–3·3 times as long as the calyx, 4–5·2 mm. long and 1·8–2·5 mm. in diam., glabrous outside, inside around the anthers with a small pilose zone and futhermore glabrous; lobes 0·4–0·8 times as long as the tube, obliquely obovate to nearly hook-shaped, curved to the right, 2·5–4 × 2–4 mm., entire, glabrous on both sides. Stamens inserted 1·5–3 mm. below the mouth of the corolla; anthers 2·1–2·7 × 0·8–1 mm. Pistil 2·5–5 mm. long; ovary 0·8–1·3 × 1–1·2 × 0·5–1·5 mm., glabrous or with a few papils; style 1–3 × 0·2 mm., glabrous; clavuncula 0·4–0·9 × 0·3–0·6 mm.; stigma c. 0·2 mm. long; ovules 8–15 in each. Fruits yellow to

Tab. 103. SCHIZOZYGIA COFFAEOIDES. 1, flowering branch (×⅔), from *Peter* 58268; 2, leaf (×⅔), from *Stolz* 1693; 3, flower (×6); 4, opened corolla (×6); 5, part of calyx with pistil (×6), 3–5 from *Peter* 58268; 6, fruits (×1), from *Bamps* 6301; 7, transverse section of fruit (×4), from *Boivinannis* 1847–1852; 8–9, seed both sides (×4), 8–10 from *Bamps* 6301.

orange; carpels ellipsoid, laterally compressed 7–15 × 5–10 × 3–5 mm., dehiscent along an adaxial line of dehiscence; wall glabrous. Seeds dark brown; 5–6 × 3–5 × 2–3 mm.; embryo white.

Malawi. N: Nkhata Bay Distr., Sanga, fl. 17.xii.1972, *Pawek* 6102 (SRGH; UC).
Also known from Angola, Zaire, Somalia, Kenya, Tanzania and the Comoro Islands. Forest understorey or open woodland, on sandy or loamy soils. 0–1500 m. Flowering and fruiting throughout the year.

15. CALLICHILIA Stapf

Callichilia Stapf in F.T.A. **4**, 1: 130 (1902). — Pichon in Mém. Mus. Nation. Hist. Nat. Paris N., Sér. **27**: 224 (1948). — Beentje in Meded. Landb. Wag. **78**–7: 2 (1978).
Ephippiocarpa Markgraf in Notizbl. Bot. Gart. Berlin **74**: 310 (1923).
Hedranthera Pichon in Mém. Mus. Nation. Hist. Nat. Paris, N.S. **27**: 225 (1948).

Shrubs or lianas, glabrous, with white latex present; branches unarmed, lenticellate; trunk dichotomously branched. Leaves decussate, those of a pair equal or unequal, subsessile or petiolate; petiole with colleters at the base. Inflorescences 1–2 in the forks of branches, pendulous, pedunculate, congested, cymose; bracts with colleters. Flowers actinomorphic, often fragrant. Sepals connate at the extreme base only, imbricate, much shorter than the corolla, inside with 1–4 rows of minute glands near the base. Corolla tube carnose, glabrous except for pilose ridges leading downward from the base of the filaments inside; tube cylindrical in the lower part, wider in the upper and there cylindrical or infundibuliform; lobes membranaceous, obtriangular, oblique, with two apices, one acute and one rounded. Stamens included, connivent in a cone. Filaments shorter than the anthers; anthers glabrous, auriculate at the base, 2-celled. Pistil glabrous; carpels connate at the very base only; disk present; style slender, clavuncula 5-winged; stigma small. Fruit berry-like, many-seeded, the two mericarps free, or (only in *C. orientalis*) semi-syncarpous; wall thin. Seeds ovoid, testa reticulate and deeply pitted; endosperm copious, surrounding the straight embryo.

An African genus of 7 species.

Callichilia orientalis S. Moore in Journ. Linn. Soc., Bot. **40**: 139 (1911). — Beentje in Meded. Landb. Wag. **78**–7: 22 (1978). TAB. **104**. Type: Mozambique, Boca, lower Búzi R. bank, *Swynnerton* 1148 (BM, holotype; K, isotype).
Ephippiocarpa orientalis (S. Moore) Markgraf in Notizbl. Bot. Gart. Berl. **74**: 310 (1923). Type as above.
Conopharyngia humilis Chiov. in Atti Soc. Nat. Mat. Modena **66**: 10 (1935). Type from Somalia.
Ephippiocarpa humilis (Chiov.) Boiteau in Adansonia, Sér. 2, **16**, 2: 280 (1976). Type as above.

Erect shrub, 1–3 m. high. Leaves petiolate; petiole 4–11 mm. long; lamina thinly coriaceous, medium green and shiny above, paler beneath, narrowly ovate, 2·5–4 times as long as wide, 4–12 × 1–4 cm., acute to acuminate at the apex (acumen 5–15 mm. long), cuneate at the base; secondary veins 8–16, tertiary venation conspicuous. Inflorescence solitary; pedicels 7–19 mm. long. Flowers fragrant. Sepals elliptic, rounded at apex, 4–9 × 2–4 mm.; calyx reflexed in fruit. Corolla white, tube 13–21 mm. long and widening at ½–⅔ of its length from the base; lobes overlapping to the left, 8–18 × 7–15 mm. Anthers 2·8–3·5 × 0·6–0·9 mm., obtuse at the apex. Ovary 1·8–1·9 × 1·4–1·7 mm.; style 8·4–11 mm. long; clavuncula 0·6–0·9 mm. high; stigma 0·4–0·9 mm. high. Fruit with the two mericarps syncarpous for ⅔–⁹⁄₁₀ of their length, smooth but for a few ridges towards the apex, 5–28-seeded; seeds 5–7 × 2·5–3·5 mm.

Mozambique. MS: Sofala Prov., Boca, on lower Búzi R. bank, fl. xii, *Swynnerton* 1148 (BM, holotype; K, isotype). GI: Inhambane Prov., south of Save R. between Morrumbene and Massinga, fl. & fr. 26.ii.1955, *E.M. & W.* 659 (LISC; SRGH).
Also in Somalia and SE. Africa. Coastal forests.

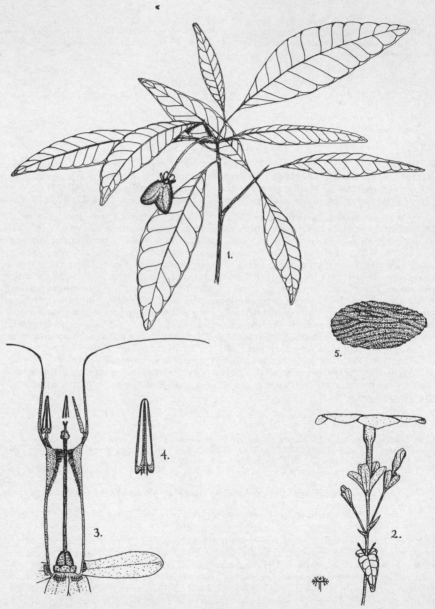

Tab. 104. CALLICHILIA ORIENTALIS. 1, fruiting branch ($\times\frac{1}{2}$), from *Vahrmeyer* 434; 2, inflorescence showing one fully opened flower ($\times 1$); 3, flower in longitudinal section ($\times 3$); 4, stamen ($\times 6$), 2–4 from *Tinley* 317; 5, fruit ($\times 4$), from *Tinley* 213.

16. DIPLORHYNCHUS Welw. ex Ficalho & Hiern

Diplorhynchus Welw. ex Ficalho & Hiern in Trans. Linn. Soc., Ser. 2, Bot. **2**: 22 (1881).

Neurolobium Baill. in Bull. Soc. Linn., Paris **1**: 749 (1881).

Shrub or tree with white or yellow latex. Branchlets usually dotted with paler lenticels. Leaves opposite or rarely subopposite, those of a pair equal, often fairly long-petiolate. Inflorescence thyrsoid, terminal and in axils of the upper leaves, lax to congested; bracts obscure. Flowers actinomorphic except for the sometimes slightly unequal sepals. Sepals ovate, acute, connate at the base. Corolla hypocrateriform; tube usually much longer than the calyx, nearly cylindrical, ventricose in the upper half and constricted at the mouth, inside velutinous to sericeous except for the glabrous base; lobes narrowly oblong or narrowly obovate, inside at the base pilose or sericeous, overlapping to the left in bud, spreading. Stamens included; inserted near the middle of the corolla tube; filaments filiform, much widened at the apex; anthers triangular, completely fertile, mucronate, sagittate at the base. Ovary of two basally connate carpels; 4 ovules on an adaxial placenta in each carpel; clavuncula subcylindrical. Fruit composed of two obliquely oblong follicles, outside dotted with many paler lenticels, dehiscent at the adaxial side. Seeds laterally compressed, obliquely oblong, long-winged at the apex; wings of the two outer seeds directed towards the base and those of the two central seeds towards the apex of the fruit.

A monotypic genus in tropical and southern Africa.

Diplorhynchus condylocarpon (Muell. Arg.) Pichon in Mém. Mus. Nation. Hist. Nat. Paris, Sér. 2, **19**: 368 (1947). — Codd in Fl. Southern Afr. **26**: 265 (1963). — Plaizier in Meded. Landb. Wag. **80**–12: 28, fig. 6, map 6 (1980). TAB. **105**. Type probably from Angola.

    Aspidosperma? condylocarpon Muell. Arg. in Martius, Fl. Bras. **6**, 1: 55 (1860). Type as above.

    Neurolobium cymosum Baill., loc. cit. Type as above.

    Diplorhynchus mossambicensis Benth. ex Oliv. in Hooker, Ic. Pl. **14**: t. 1355 (1881). — Stapf in F.T.A. **4**, 1: 107 (1902). Type: Malawi, Shire Highlands, *Buchanan* s.n. (K, lectotype).

    Diplorhynchus psilopus Welw. ex Ficalho & Hiern, tom. cit. 23. — Stapf in F.T.A. **4**, 1: 106 (1902). Type from Angola.

    Diplorhynchus angolensis Büttner in Verh. Bot. Ver. Prov. Brand. **31**: 85 (1890). — Stapf, loc. cit. Type from Angola.

    Diplorhynchus welwitschii Rolfe in Bol. Soc. Brot. **11**: 85 (1893). — Stapf, tom. cit.: 105 (1902). Type from Angola.

    Diplorhynchus poggei K. Schum. in Engler & Prantl, Nat. Pflanzenfam. **4**, 2: 142, fig. 54 O (1895). Type from Angola.

    Diplorhynchus angustifolia Stapf, tom. cit.: 107 (1902). Type from Tanzania.

    Diplorhynchus condylocarpon subsp. *mossambicensis* (Benth. ex Oliv.) Duvign. in Bull. Soc. Roy. Bot. Belg. **84**: 265 (1952). Type as for *D. mossambicensis*.

    Diplorhynchus condylocarpon subsp. *mossambicensis* var. *mossambicensis* f. *angustifolia* (Stapf) Duvign., loc. cit. Type as for *D. angustifolia*.

    Diplorhynchus condylocarpon subsp. *mossambicensis* var. *psilopus* (Welw. ex Hiern & Ficalho) Duvign., loc. cit. Type as for *D. psilopus*.

    Diplorhynchus condylocarpon subsp. *mossambicensis* var. *psilopus* f. *microphylla* Duvign., tom. cit.: 266 (1952). Type from S. Africa.

    Diplorhynchus condylocarpon subsp. *angolensis* (Büttner) Duvign., loc. cit. Type as for *D. angolensis*.

A tree or shrub, sometimes scandent, 1–many-stemmed, (1)3–12(20) m. tall. Trunk 0·10–0·50(2·0) m. in diam.; bark smooth to rough, usually longitudinally fissured or reticulate, greyish, brownish to blackish; wood whitish yellow to pale orange. Branchlets drooping, puberulent to glabrous. Leaves petiolate; petiole (0·5)1–2(3·7) cm., glabrescent to puberulent, above sometimes tomentose, and there with or without a scale-like colleter and/or 1–4 rows of 1–4 smaller glands; lamina variable in shape, obovate, elliptic, subcircular or ovate, (1·1)1·5–2·4(3·2) times as long as wide, (2·6)3·8–9·3(12·1) × (1·1)2·2–5·4(6·7) cm., acute, rounded to emarginate and acuminate to mucronate at the apex, cuneate to obtuse at the base, flat to undulate or sometimes somewhat crispate, thinly coriaceous to leathery, glabrous to pubescent, sometimes only puberulent on the margins, costa puberulent (sometimes the basal part only), rarely glabrous; secondary veins (6)8–14(19) pairs, conspicuous, glabrous or somewhat puberulent, sometimes with axillary tufts of glandular hairs;

Tab. 105. DIPLORHYNCHUS CONDYLOCARPON. 1, flowering branch ($\times\frac{2}{3}$), from *Norrgrann* 248;
2, 3, 4, leaf habit ($\times\frac{2}{3}$), 2 from *Brass* 17414, 3 from *Swynnerton* 39, 4 from *Schlieben* 3036; 5,
group of flowers ($\times 4$), from *Schlieben* 3036; 6, petals, above ($\times 4$); 7, anthers and pistil
($\times 6$); 8, pistil ($\times 6$), 6–8 from *Norrgrann* 248; 9, fruits ($\times\frac{2}{3}$), from *Brass* 17414; 10, fruit,
open with two seeds ($\times\frac{2}{3}$), from *Schlieben* 3036.

tertiary veins inconspicuous. Inflorescence (1·5)2–9(14) × (2)2·5–9(13) cm.; bracts obscure, rounded, glabrous to pubescent, sometimes with glandular hairs. Peduncle 0·6–4(4·5) cm. long; pedicels glabrous to tomentose, sometimes with glandular hairs, 0·5–2·5(3·0) mm. long. Flowers very sweet-scented. Sepals pale green or yellow-green, sometimes slightly unequal, (0·5)1–2(3) times as long as wide, (0·5)0·75–1(1·5) × (0·25)0·5(0·75) mm., outside pubescent to tomentose at the base, pubescent to glabrous towards the apex, very rarely entirely glabrous, glabrous to appressed-puberulent inside especially towards the apex; margins entire, usually hyaline, scabrid to hirsute, rarely glabrous; receptacle pubescent to tomentose, rarely glabrous to puberulent, with glandular hairs. Corolla white to creamy, very rarely reddish to orange, tube (1·5)2–5(7) times as long as the calyx, (1·5)2–3(3·2) × 1–1·5(1·8) mm., glabrous to slightly puberulent outside, lobes (3·5)5(6) × (0·7)1·2(2) mm., rounded or slightly acute, entire, glabrous to puberulent outside, usually with many glandular hairs; buds tinged red; with a scale between the bases of all lobes (scale glabrous, (0·5)0·7–1(1·2) mm. long, and sometimes at the base with some long scabrid hairs, and connected by its edges with the lobes). Stamens with the filaments usually scabrid to hispid, rarely glabrous, 0·2–0·5(0·8) mm. long; anthers 1–1·2(1·5) × 0·5 mm., with a 0·1–0·2(0·5) mm. long mucro. Pistil (1·2)1·7–2·2(2·4) mm. long; ovary (0·3)0·4(0·5) × (0·3)0·5(0·8) × (0·5)0·8(1·0) mm., of two carpels, rounded at the apex, glabrous or slightly puberulous; style subapical, filiform, (0·5)0·7–1·3 × 0·2–0·3 mm., not split at the base, glabrous to very slightly puberulous; clavuncula lanuginose, sometimes glabrescent towards the base, (0·2)0·3–0·4 × 0·3–0·4 mm.; apiculum bifid, (0·2)0·3–0·4(0·7) mm. long, glabrous. Fruit composed of two follicles, which are diverging by 180°, woody, green or pale to dark brown, 1·8–3·3(4·2) times as long as wide, (2·2)2·9–6·6 × 1·1–2·2 × 0·4–1·1(1·8) cm., coherent at the base, obliquely oblong, glabrous or slightly puberulent, 2-valved, each valve 2-seeded. Seed (2·5)3·5–4·5(5·5) cm. long; grain dark brown, elliptic, (1)1·3–2·0 × 0·7–1·2(1·8) cm., wing diaphanous, with longitudinal veins, 2·0–2·5(3·2) × 1–1·5(1·9) cm.

Botswana. N: Chobe National Park, Gubatsa Hills, fl. 23.x.1972, *Pope, Biegel & Russell* 836 (K; LISC; PRE; SRGH). **Zimbabwe.** N: Lomagundi Distr., Mutorashanga (Mtoroshanga) Pass, *Rodin* 4417 (K; MO; NY; S; SRGH; UC; US; WU). W: Hwange (Wankie), *Levy* 1120 (E; K; MO; PRE; SRGH). C: Marondera (Marandellas), *West* 3110 (K; SRGH). E: Chimani-mani (Melsetter) Distr., Rusitu R., *Barrett* 95/56 (COI; LISC; MO; SRGH). S: Mberengwa (Belingwe) Tribal Trust Land, 32 km. N.E. of West Nicholson, fr. v.1956 *Judge* 5/56 (K; LISC; MO; SRGH). **Malawi.** N: Rumphi Distr., 4·5 km. West of Rumphi, 15.iv.1975, 3500 ft., *Pawek* 9202 (K; SRGH; UC). C: Kasungu, 25.viii.1946, 1000 m., *Brass* 17414 (BR; K; MO; NY; PRE; SRGH; UC). S: Chikwawa, *Brass* 17987 (K; MO; NY; SRGH; US). **Mozambique.** N: Erati, Namapa, road to Mirrote near aerodrome, fr. (immat.) 15.iii.1960, *Lemos & Macuácua* 41 (BM; COI; K; LISC; PRE; SRGH). Z: Mocuba Distr., Namagoa Plantations, *Faulkner* 178 (BR; G; LISJC; PRE; S; SRGH). T: Mágoè, *Chase* 2720 (BM; NY; SRGH).
Also in Angola, southern Zaire, Tanzania, Namibia, Botswana and S. Africa. Widespread in dry deciduous woodland and stony hillsides, somewhat tolerant of the toxic serpentine-derived soils of the Great-Dyke in Zimbabwe where it becomes more common. Alt.: 0–1700 m.

17. CATHARANTHUS G. Don

Catharanthus G. Don, Gen. Syst. **4**: 95 (1837). — Codd in Fl. Southern Afr. **26**: 267 (1963). — Markgraf in Fl. Madag., Apocynac.: 140 (1976). — Plaizier in Meded. Landb. Wag. **81**–9: 2 (1981).
Lochnera Rchb., Consp. **1**: 134 (1828), nom. nud.; Rchb. ex Endl., Gen. Pl.: 583 (1838). — Stapf in F.T.A. **4**, 1: 118 (1902).

Herbs, perennial or annual, often woody at the base, branched to a varying degree. Leaves herbaceous to thinly coriaceous, opposite, mucronate, with a fringe or intra- and interpetiolar colleters. Inflorescences terminal, but apparently lateral due to pseudomonopodial continuation of the stem by alternating development of one of the axillary buds of the apical pair of leaves, 1–2-flowered. Flowers actinomorphic. Sepals narrowly to very narrowly oblong, subulate, eglandular. Corolla purple, red, pink, or white, hypocrateriform; tube laxly puberulous or glabrous, constricted and woolly to velvety at the throat; lobes spreading, obliquely obovate, spiculate. Stamens inserted at the widest portion of the corolla-tube (usually above the middle); anthers oblong obtuse at the base. Ovary of two very narrowly oblong carpels; disk composed of two narrowly triangular to narrowly oblong glands, the bases of which touch each other at the abaxial sides of the carpels; style filiform; clavuncula cylindrical, at the base provided with a reflexed hyaline frill; ovules numerous. Fruit composed of two follicles; follicles cylindrical, acute. Seed black, oblong; testa rugose; hilum lateral; cotyledons flat, shorter than the radicle; endosperm scanty.

A genus of 8 species, one restricted to India and Sri Lanka, all the others confined to Madagascar. Of the latter group one, *C. roseus* (L.) G. Don cultivated and naturalized all over the tropics, especially in coastal areas.

Catharanthus roseus (L.) G. Don, Gen. Syst. **4**: 95 (1837). — Pichon in Mém. Mus. Nation. Hist. Nat. Paris, Sér. 2, **27**: 238 (1948). — Codd in Fl. Southern Afr. **26**: 267 (1963). — Stearn in Lloydia **29**: 196 (1966); in W. T. Taylor & N. R. Farnworth, Syn. Catharanthus 35 (1975). — Markgraf in Fl. Madag., Apocynac.: 152 (1976), partly excl. syn. *Hottonia litoralis* Lour. — Plaizier in Meded. Landb. Wag. **81**–9: 3, fig. 1, phot. 1 (1981). TAB. **106**. Type: Miller, Fig. Pl. Gard. Dict. **2**: t. 186 (1757) (Lectotype).
   Vinca rosea L., Syst. Nat. ed., 10: 944 (1759). Type as above.
   Pervinca rosea (L.) Moench., Method. 463 (1794). Type as above.
   Vinca speciosa Salisb., Prodr.: 147 (1796). Type as above.
   Lochnera rosea (L.) Rchb., Consp.: 134 (1828); Rchb. ex Endl., Gen. Pl.: 583 (1838). — Stapf in F.T.A. **4**, 1: 118 (1902). Type as above.
   Catharanthus roseus var. *nanus* Markgraf in Adansonia, Ser. 2, **12**: 222 (1972). Type from Madagascar.

Suffrutex 30–100(–200) cm. high, erect or decumbent, usually with white latex. Stems ± terete, green or yellowish-green, sometimes slightly to heavily suffused with red or purple, laxly pubescent or glabrous. Leaves decussate, petiolate; petiole (0·1)0·3–1 cm. long, laxly puberulous or glabrous, with a fringe of colleters in the axil, the outer ones longer than the inner and with some strigose hairs; lamina rather variable in shape, elliptic, obovate or narrowly obovate, 1·9–3 times as long as wide, (3)4–9 × (0·8)1·5–3·5 cm., obtuse or acute, with a mucronate apex and sometimes slightly emarginate, cuneate or obliquely cuneate at the base, laxly pubescent to glabrous on both sides; veins paler; secondary veins more or less conspicuous, 7–11 on each side; tertiary venation inconspicuous. Inflorescence ebracteate. Pedicel 0·1–0·2 mm. long, laxly puberulous to glabrous. Flowers (3)4–5(5·6) cm. long. Sepals green, slightly connate at the base, sometimes slightly unequal, 2·7–4·7 times as long as wide, (2)3–5 × 1–1·5 mm., outside laxly puberulous or glabrous, glabrous inside, sometimes towards the apex with some minute white hairs, entire erect. Corolla white or pink, with a purple, red, pink, pale yellow or — if white sometimes — white centre; tube often slightly greenish, with a long narrow cylindrical basal and a short wider upper portion (4·2)5–8(12·5) times as long as the calyx, 2·2–3 cm. long (in the narrow portion 1–2 mm. wide, up to 3 mm. in the widened portion), outside laxly puberulous or glabrescent, within the throat at the level of the anthers with a c. 0·5 mm. broad densely strigose ring, and then in the widened portion velutinous for 1–1·5 mm. on the veins from the base of the filaments, and below it with a 1·5–2 mm. broad sericeous ring, situated just below the level of the clavuncula; throat 1·5–2 mm. in diam.; lobes usually paler outside, (0·8)1·2–2(2·8) × (0·6)0·9–2 cm. broadly obovate, spreading, entire, sometimes ciliate, in bud overlapping to the left. Stamens included, inserted 0·4–0·6 cm. below the corolla-mouth; filaments very short, glabrous; anthers with the apex acute, cordate at the base, 2·3–3 × 1–1·3 mm., introrse, completely fertile. Pistil 17–26 mm. long; carpels 1·5–2(2·8) × 0·8–1 × 0·5–1 mm., connate at the base, rounded at the apex, puberulous at the apex and glabrous towards the base; style 16–24 × 0·3 mm., sometimes slightly split at the base, glabrous; clavuncula 1·3–2·3 mm. long, at the apex with a woolly ring, (0·1)0·3 × 0·8–1·3 mm., at the base also with a woolly ring, 0·3–0·5 × (0·8)1–1·5 mm., and in between with a glabrous or puberulous 0·4–0·8 × 0·5–1 mm. wide zone, and at the base with a reflexed hyaline frill, 0·5–1 × 0·8–1·5 mm. Fruit green, also when mature, composed of two follicles (sometimes one aborted or reduced) erect or slightly spreading. Follicle cylindrical, striate, 5·2–13·5 times as long as wide, 1·2–3·8 × 0·2–0·3 cm., dehiscent at the adaxial side. Seeds numerous grooved at one side, 1–2 × 0·5–0·8 mm.

Zambia. C: Lusaka, fl. vii, *King* 47 (K). **Malawi.** N: 8 km. E. of Mzuzu, fl., fr. 12.iii.1978 *Phillips* 3321 (WAG); S: Kundwelo, fl. & fr. 29.vii.1956 *Newman & Whitmore* 278 (BR, SRGH, WAG). **Mozambique.** N: Nampula, Ríbauè, fl., fr. 21.iii.1964 *Correia* 217 (LISC).

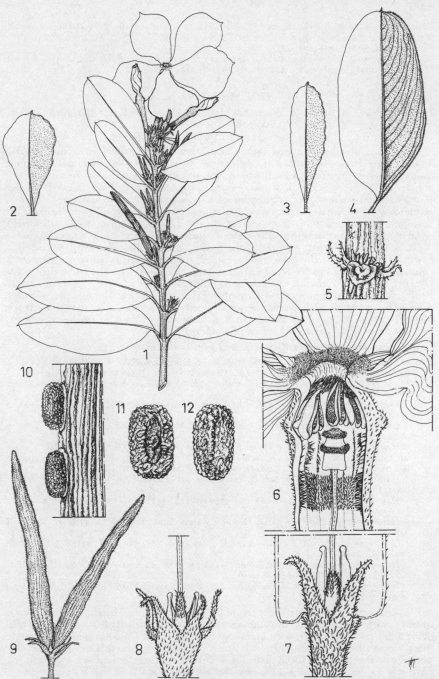

Tab. 106. CATHARANTHUS ROSEUS. 1, habit ($\times\frac{2}{3}$), from *van Meer* 627; 2, 3, 4, leaves ($\times\frac{2}{3}$), 2–3 from *Leeuwenberg* 10887, 4 from *Wild* 4423; 5, colleters in leaf axil ($\times 6$), from *Schlieben* 2653; 6, apical portion of the corolla-tube and apex of pistil ($\times 6$), from *Stubbings* 198; 7–8, calyx with pistil base and disk-glands ($\times 6$), 7 from *Stubbings* 198, 8, from *Hunziker* 11; 9, fruits ($\times 1\frac{1}{2}$), from *Wild* 4423; 10, seeds on the placenta ($\times 8$), from *Decary* 18858; 11, seed, adaxial ($\times 8$), 12, seed, abaxial ($\times 8$), 11–12 from *Wild* 4423.

Z: Zambézia, fl., fr. 25.ii.1912 *Rogers* 4559 (SRGH). MS: Andrada, fl., fr. 25.xi.1904 *Vasse* 120 (P). GI: Inhambane Prov., fl. & fr. 30.x.1938 *Torre* 1589 (COI, LISC). M: Inhaca Isl. fl. & fr. 8.vii.1956 *Mogg* 31702 (K, SRGH).

Indigenous to Madagascar but cultivated and naturalized throughout the Tropics of both hemispheres and sometimes extending to the subtropics.

18. HOLARRHENA R. Br.

Holarrhena R. Br. in Mem. Wern. Soc. **1**: 62 (1811). — De Kruif in Meded. Landb. Wag. **81**–2: 4 (1981).

Physetobasis Hassk. in Flora **7**: 104 (1857).

Deciduous shrubs or trees with white latex and unarmed branches. Leaves opposite or subopposite, those of a pair equal, shortly petiolate, without colleters in the axils; petiole glandular near the base. Stipules intrapetiolate and often obscure or absent. Inflorescences terminal and/or apparently axillary, cymose, many-flowered. Flowers sweet-scented, actinomorphic or with sepals only unequal. Sepals free or connate at the very base, subequal or sometimes unequal, with colleters inside at the base and alternating with the sepals. Corolla hypocrateriform; tube cylindrical, slender, wider near the base and apex; lobes contorted in bud and overlapping to the right. Stamens included, inserted at the base of the corolla-tube; anthers glabrous, basifixed, rounded at the base, mucronate at the apex. Ovary superior, carpels 2, connate at the very base only, ovoid, gradually narrowed into the style; disk none; style cylindrical; clavuncula ovoid; stigma bifid; placentas adaxial; ovules numerous. Fruit composed of 2 scarcely diverging mericarps; mericarps follicular, slender, more or less constricted between the seeds. Seeds numerous, linear to very narrowly oblong, striate, grooved, glabrous, with a dense tuft of hairs at the apex; embryo large, surrounded by the very scanty endosperm.

An African and Asian genus of 4 species, of which 2 occur in Africa.

Holarrhena pubescens (Buch.-Ham.) Wall. ex G. Don, Gen. Syst. **4**: 78 (1837). — Codd in Fl. Southern Afr. **26**: 263 (1963). — K. Coates Palgrave, Trees of Southern Afr.: 786, pl. 262 (1977). — De Kruif in Meded. Landb. Wag. **81**–2: 17, fig. 5–6, map 4–5 (1981). TAB. **107**. Lectotype from Burma.

Echites pubescens Buch. — Ham. in Trans. Linn. Soc. **13**: 521 (1822), non Hook. & Arnott (1841). Type as above.

Chonemorpha pubescens (Buch.-Ham.) Sweet, Hort. Brit., ed. **3**: 458 (1839). Type as above.

Holarrhena febrifuga Klotzsch in Peters, Reise Mossamb., Bot. **1**: 277 (1862). Type: Mozambique; Sena, *Peters* s.n. (holotype B†); 16°55′S, 33°E, 5 km. E. of Nyamapanda, 600 m., *Gillett* 17510 (K, L, SRGH, isoneotypes; WAG, neotype).

Holarrhena tettensis Klotzsch in Peters, op. cit. **1**: 278 (1862). Type: Mozambique, Sena, *Peters* s.n. (B†).

Holarrhena glabra Klotzsch in Peters, op. cit. **1**: 279 (1862). Type: Mozambique, Tete, *Peters* s.n. (B†).

Holarrhena febrifuga var. *glabra* Oliver in Trans. Linn. Soc. **29**: 108 (1875). Type from Tanzania.

Holarrhena fischeri K. Schum. in Engl., Pflanzenw. Ost-Afr., **C**: 316 (1895). Lectotype from Tanzania.

Holarrhena febrifuga f. *grandiflora* Stapf ex De Wild. in Ann. Mus. Congo Belge, Bot., Sér. **4**, 1: 101 (1903). Type from Zaire.

Holarrhena glaberrima Markgraf in Mitt. Bot. Staatss. München **1**: 28 (1950). Type from Tanzania.

Shrub or tree, 0·60–18 m. high; trunk 12–25 cm. in diam.; bark smooth and more or less distinctly lenticellate or rough and corky, longitudinally fissured, pale to dark grey; inner bark pale buff, with green outer layer, wood soft. Branches lenticellate; branchlets pubescent or less often glabrous. Leaves very variable in shape and size; petiole pubescent or less often glabrous, 0·2–12 mm. long; lamina ovate to elliptic or narrowly so, 1·7–19·5 × 1·3–11·2 cm., acuminate or acute at the apex, cuneate or rounded at the base, pubescent to glabrous on either side: if hairy usually more so beneath; lamina of the leaves of the first pair or the first two pairs of a branchlet often subcircular, rounded or obtuse at the apex and smaller; on each side of the midrib with 5–25 secondary veins. Inflorescences usually apparently axillary or leaf-opposed, sometimes terminal, fairly lax. Peduncle, branches and pedicels pubescent or less often glabrous; peduncle 0·9–1·7 cm. long; bracts linear, pubescent or less often glabrous on both sides. Sepals (1)2–12 × 0·5–1·7 mm., elliptic to linear, pubescent outside, often only towards the apex, indumentum a mixture of

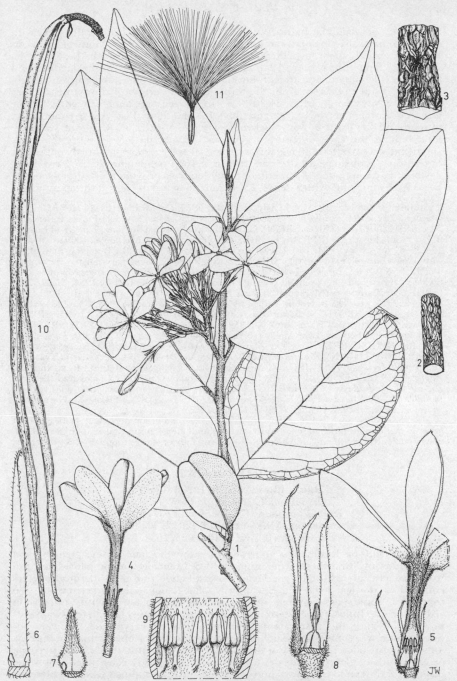

Tab. 107. HOLARRHENA PUBESCENS. 1, flowering branch ($\times \frac{2}{3}$), from *Pawek* 7734; 2, part of branch with lenticels ($\times \frac{2}{3}$), from *Thulin & Mhoro* 1256; 3, bark ($\times \frac{2}{3}$), 4, flower ($\times 2$), 3–4 from *Leeuwenberg* 10867; 5, part of flower showing stamens and pistil ($\times 2$), 6, sepal inside with colleters ($\times 6$), 5–6 from *Leeuwenberg* 10866; 7, sepal inside with colleter ($\times 8$), from *Squires* 778; 8, part of calyx with sepals ($\times 4$), 9, part of corolla with stamens ($\times 8$), 8–9 from *Leeuwenberg* 10866; 10, fruit ($\times \frac{2}{3}$), from *Lawton* 1212; 11, seed ($\times \frac{2}{3}$), from *Harmand* 23.

eglandular and glandular hairs, less often entirely glabrous, inside pubescent with eglandular hairs all over or only at the apex and then the basal 0·5–3 mm. glabrous, or less often entirely glabrous, ciliate, with 0–10 colleters inside at the base. Corolla 20–38 mm. long; tube white 9–19 mm. long, pubescent outside with eglandular and also some glandular hairs except for the basal 0·5–3 mm., sometimes entirely glabrous, inside pubescent from the throat to the insertion of the stamens; lobes white, narrowly triangular, 10–24(30) × 3–8 mm., recurved, pubescent and ciliate or glabrous. Anthers narrowly triangular, 0·8–1·8 mm. long. Pistil 1·8–3·1 mm. long; ovary 0·7–1·2 × 0·5–1·0 mm. glabrous; style 0·3–1·2 × 0·2–0·3 mm., shortly pubescent at the apex or entirely glabrous; clavuncula 0·3–1·2 × 0·2–0·5 mm., shortly pubescent or less often glabrous; stigma 0·1–0·3 mm. long, shortly pubescent or glabrous. Fruits pale grey to dark brown; carpels pendulous, 20–37·5 cm. long, 2–9 mm. in diam., outside sometimes many lenticels or white-dotted or -spotted. Seeds 9–16 mm. long, with a dense, 25–45 mm. long tuft of hairs at the apex.

Zambia. B: Zambezi (Balovale) Distr., fr. 8–10.v.1953, *Holmes* 1880 (FHO; K; SRGH). N: Chiengi, fl. 12.x.1949, *Bullock* 1244 (BR; K). W: Ndola Distr., Mindola Forest Reserve, fl. 20.xii.1951, *Holmes* 494 (FHO; SRGH). C: Mt. Makulu Research Station, 19 km. S. of Lusaka, fl. 7.xi.1956, *Angus* 1439 (BM; BR; K; PRE). E: Jumbe, Chikowa Mission, c. 1000 m., fl. & fr. 13.x.1958, *Robson* 81 (BM; BR; LISC; PRE; SRGH). S: Mazabuka Research Station, fr. 21.iv.1965, *Lawton* 1212 (FHO; SRGH). **Zimbabwe.** N: Centenary Distr., Mzarabani T.T.L., fr. 9.iv.1965, *Bingham* 1449 (BR; LISC; M; PRE; SRGH). W: Eastern Matopos, near Switsha Lumen Falls, between old Gwanda Road and Mtshabezi, fr. 16.v.1954, *Plowes* 1726 (SRGH). C: Marondera (Marandellas) Distr., between Dabi R. and Macheke R., fl. 8.i.1950, *Munch* 222 (K; SRGH). E: Honde Valley, Mtarazi Falls, fl. & fr. 5.xi.1948, *Chase* 1319 (BM; COI; K; LISC; SRGH). S: Beitbridge, 1200 m., fl. xi.1952, *Davies* 377 (MO; SRGH). **Malawi.** N: 40 km. S. of Karonga, fl. 9.i.1959, *Robinson* 3128 (K; M; MO; PRE; SRGH). C: Senga Bay Hotel, between Lake Malawi (Nyasa) and hotel, 480 m., fr. 17.ii.1959, *Robson* 1639 (BM; BR; K; LISC; PRE; SRGH). S: Mangochi (Fort Johnston), fl. & fr. 12.xi.1954, *Jackson* 1378 (BM; BR; FHO; K). **Mozambique.** N: Metangula, at edge of Lake Malawi (Nyasa), fl. 10.x.1942, *Mendonça* 738 (LISC). Z: Mocuba Distr., Namagoa, 60–120 m., fl. xi–xii.1944, *Faulkner* 328 (BR; G; K; LD; PRE; SRGH). T: Maravia, between Chicoa and Tete, fr. 25.vi.1949, *Andrada* 1643 (COI; LISC). MS: Sofala Prov., near Gorongosa (Vila Paiva de Andrada), 350 m., fl. 22.x.1961, *Gomes e Sousa* 4733 (COI; K; LISC; PRE). GI: Gaza Prov., Massangena, fr. vii.1932, *Smuts* 358 (BM; K; NY; PRE).

Also in Zaire, Kenya, Tanzania, S. Africa (Transvaal) and in southern Asia. Widespread at lower to medium altitudes in medium to lower rainfall areas, in mixed deciduous woodland, open woodland or scrubland or in bamboo forest. In sandy or stony ground or rocky outcrops.

19. RAUVOLFIA L.

Rauvolfia L., Sp. Pl. **1**: 208 (1753).
Ophioxylon L., Sp. Pl. **1**: 1043 (1753).
Dissolaena Lour., Fl. Cochinch.: 137 (1790).
Cyrtosiphonia Miq. in Fl. Ind. Bat. **2**: 401 (1857).
Heurckia Muell. Arg. in Flora **53**: 168 (1870).

Trees, shrubs or rhizomatous undershrubs with white latex. Spines, tendrils and stipules absent. Branches terminating in 0–5 branchlets accompanied by 5–0 inflorescences; branch-nodes marked by a distinctive ring of numerous colleters confined to the leaf axils and the interpetiolar region. Leaves verticillate, 3–6 together, in some species on some nodes opposite, those of a whorl often unequal in size and shape. Inflorescences terminal, sometimes seemingly lateral, few- to many-flowered, cymes; the first ramification umbellate or corymbose. Flowers actinomorphic or in some species slightly zygomorphic. Calyx lobes connate at the base only, without colleters inside. Corolla hypocrateriform, infundibuliform, urceolate or campanulate, glabrous outside; tube cylindrical, slightly constricted below the insertion of the stamens; lobes in bud contorted and overlapping to the left. Stamens inserted above the middle of the corolla tube; filaments short; anthers free from each other and the pistil, included, glabrous, basifixed, cordate at the base, mucronate at the apex. Ovary superior, composed of two, free or partly to completely fused carpels and with 1–2 ovules; disk annular or cyathiform, entire or lobed; style cylindrical; clavuncula conspicuous, bi-apiculate. Fruits apocarpous or partly to completely syncarpous drupes; often only 1 carpel developing; each carpel with a single stone containing a single seed. Seeds laterally compressed, obliquely ovate or elliptic; embryo large, surrounded by a thin carnous endosperm.

A pantropical genus of c. 60 species, 8 of which indigenous in Africa and Madagascar.

1. Leaf-lamina linear to narrowly obovate, 2–9 × 0·2–2 cm., with 4–9 secondary veins on each side of the midrib. Inflorescences 3–6-flowered. Peduncles very short. Rhizomatous undershrubs, 10–30 cm. high - - - - - - - - - - - 1. *nana*
– Leaf-lamina narrowly ovate to narrowly obovate, often larger, 1·8–50(70) × 0·7–15(19) cm., with 6–35(45) secondary veins on each side of the midrib. Inflorescences few- or many-flowered. Peduncles 0·2–15·5 cm. long. Shrubs or trees up to 40 m. high - - - 2
2. Leaf-lamina distinctly acuminate (acumen 2–35 mm. long). Secondary branches of inflorescences not longer than 2 cm. - - - - - - - - - 2. *mannii*
– Leaf-lamina subacute, less often subacuminate, but never with a long, distinct acumen 3
3. Pedicels 0·1–2 mm. long. Inflorescences congested in ultimate branchings. Fruits dark red, often lenticellate; when apocarpous only one carpel developing (sub-globose and 5–20 mm. in diam.) when syncarpous bilobed, 10–30 × 7–23 × 7–24 mm. Trees up to 40 m. high; trunks 5–100 cm. in diam. - - - - - - - - - - - 3. *caffra*
– Pedicels 3–8 mm. long. Inflorescences fairly lax. Fruits orange, yellowish- or reddish-orange, one or both carpels developing; carpels elliptical, 5–10 × 4–8 mm. Shrubs or small trees, 0·50–9 m. high; trunks up to 10 cm. in diam. - - - - - 4. *mombasiana*

1. **Rauvolfia nana** E. A. Bruce in Kew Bull. **3**: 461, pl. 2 (1949). White, F.F.N.R.: 351 (1962).
Type: Zambia, W: Mwinilunga Distr., near source of Matonchi Dambo, *Milne-Redhead* 3264 (K, holotype; BM, BR, isotypes; PRE).

Rhizomatous undershrub, 10–30 cm. high. Leaves sessile or nearly so; lamina 2–9 × 0·2–2 cm., linear to narrowly obovate, subacute to acute at the apex, at the base narrowly cuneate, papery to subcoriaceous when dried, glabrous, with 4–9 secondary veins on each side of the midrib. Inflorescences 3–6-flowered, somewhat congested, peduncle very short, glabrous. Pedicels 3–5 mm., glabrous. Sepals 1·3–1·5 mm. long, triangular, glabrous. Corolla hypocrateriform; tube white or greenish-yellow, 3·5–5 mm. long, inside pubescent from 1·4–1·8 mm. from the base to the lobes; lobes sometimes different from the tube in colour, white or greenish-yellow, 1·8–2·5 mm. long, ovate or triangular, inside pubescent at the base. Stamens inserted 2·4–3·5 mm. above the base of the corolla-tube; filaments 0·3–0·5 mm. long; anthers 1–1·2 mm. long. Pistil 2·5–3·5 mm. long; carpels c. 0·7 mm. long, glabrous; disk ring-shaped, c. 0·3 mm. high, glabrous; style c. 2 mm. long, glabrous; clavuncula 0·2–0·3 mm. long, sparsely pubescent; stigma c. 0·2 mm. long, glabrous. Fruits red with only one carpel developing; carpel ellipsoid, 7–12 × 3–8 mm. Seeds 6–12 × 3–8 × 2–5 mm.

Zambia. B: 16 km. E. of Zambezi (Balovale) on road to Kabompo R., fr. 25.iii.1961, *Drummond & Rutherford-Smith* 7327 (K; MO; PRE; SRGH; WAG). W: Mwinilunga Distr., c. 5 km. S.E. of Angola border, c. 2 6 km. S.W. of Mudileji R., c. 1300 m., fl. 6.xi.1962, *Lewis* 6151 (K; MO).
Also in Zaire and Angola. In open woodland on sandy soil.

2. **Rauvolfia mannii** Stapf in Kew Bull. **1894**: 21 (1894); in F.T.A. **4**, 1: 113 (1902). Type from Gabon.
Rauvolfia cardiocarpa K. Schum. in Engl. & Prantl, Nat. Pflanzenfam. **4**, 2: 154 fig. 56R (1895). Type from Gabon.
Rauvolfia preussii K. Schum. in Engl. & Prantl, Nat. Pflanzenfam. **4**, 2: 154 fig. 56S (1895). — Stapf, tom. cit.: 114 (1902). Type from Cameroon.
Rauvolfia obscura K. Schum. in Engl. & Prantl, Nat. Pflanzenfam. **4**, 2: 154 (1895). — Stapf tom. cit.: 117 (1902). Type from Zaire.
Rauvolfia rosea K. Schum. in Engl., Pflanzenw. Ost-Afr. **C**: 317 (1895). — Stapf, tom. cit.: 114 (1902). Type from Tanzania.
Rauvolfia longiacuminata De Wild. & T. Durand in Bull. Soc. Bot. Belg. **38**, 1: 205 (1899) ("*longeacuminata*"). — Stapf, tom. cit.: 116 (1902) ("*longeacuminata*").
Rauvolfia cumminsii Stapf, tom. cit.: 114 (1902). Addenda: 601 (1904). Type from Ghana.
Rauvolfia liberiensis Stapf, tom. cit., addenda: 601 (1904). Type from Liberia.

Understorey shrub or small tree, 0·30–8 m. high. Trunk 0·5–1 cm. in diam. or more. Bark scaly, peeling off, greenish- to grey-brown, lenticellate. Branches brown; branchlets medium green. Leaves verticillate, 3–6 together, on some nodes opposite; petiole 1–27 mm. long, glabrous; lamina 2·5–28 × 0·7–10·5 cm., narrowly ovate to narrowly obovate, acuminate at the apex (acumen 2–35 mm. long), cuneate at the base, papery to subcoriaceous, glossy, glabrous, with 6–19 secondary veins on each side of the midrib. Inflorescences few- to many-flowered; peduncle 0·2–6 cm. long, glabrous. Pedicels 1–8 mm. long, glabrous. Sepals 0·6–2·8 mm. long, triangular, glabrous. Corolla hypocrateriform; tube white, yellowish- or greenish-white, often with longitudinal pink or red stripes, 2·5–10·5 mm. long, inside pubescent from 1·5–7·5 mm. from the base to (or almost) to the apex; lobes white to pink, 0·8–3·6 mm. long, ovate to obovate, glabrous. Stamens inserted 2–8 mm. above the base of the corolla-tube; filaments 0·2–0·6 mm. long; anthers 0·8–1·5 mm. long. Pistil 1·8–6·5 mm. long; carpels 0·8–2 mm. long, glabrous; disk ring-shaped, 0·4–0·8 mm. high, glabrous; style 0·8–6 mm. long, glabrous; clavuncula 0·3–1 mm. long, often pubescent; stigma 0·1–0·6 mm. long, glabrous. Fruits red, often laterally compressed; when apocarpous one or both carpels developing; carpels ovoid to obovoid and 5–12 × 4–7 × 3–6 mm.; when syncarpous obcordate and 5–11 × 4–15 × 3–7 mm. Seeds 4–11 × 3–6 × 2–5 mm.

Malawi. N: Chitipa Distr., Misuku Hills, Mughesse Forest, fr. 2.i.1978, *Pawek* 13506 (BR). Widely distributed in tropical Africa from Liberia to Kenya in the north and from Angola to Malawi in the south. Wet places in rain-forest, riverine forest and old secondary forest.

3. **Rauvolfia caffra** Sonder in Linnaea **23**: 77 (1850). — Stapf in F.T.A. **4**, 1: 110 (1902). — Brenan et al. in Mem. New York Bot. Gard. **8**, 5: 502 (1954). White, F.F.N.R.: 351 (1962). TAB. **108**. Type from S. Africa.
  Rauvolfia natalensis Sonder, tom. cit.: 78 (1850). — Stapf, tom. cit.: 111 (1902). Type from S. Africa.
  Rauvolfia macrophylla Stapf in Kew Bull. **1894**: 20 (1894); tom. cit.: 110 (1902); — non Ruiz & Pav. in Fl. Peruv. **2**: 26, t. 152 (1799). Type from S. Africa.
  Rauvolfia ochrosioides K. Schum. in Phys. Abh. Kön. Akad. Wiss. Berlin **1**: 52 (1894); in Engl. & Prantl, Nat. Pflanzenfam. **4**, 2 154 (1895); in Engl., Pflanzenw. Ost-Afr. **A**: 88 (1895); in Engl., Pflanzenw. Ost-Afr. **C**: 318 (1895). — Stapf, tom. cit.: 111 (1902). Type from Tanzania.
  Rauvolfia inebrians K. Schum. in Engl. & Prantl, Nat. Pflanzenfam. **4**, 2: 154 (1895); in Engl., Pflanzenw. Ost-Afr. **A**: 93 (1895); in Engl., Pflanzenw. Ost-Afr. **B**: 352 (1895); in Engl. Pflanzenw. Ost-Afr. **C**: 318 (1895). — Stapf, tom. cit.: 112 (1902). Syntypes from Tanzania.
  Rauvolfia caffra var. *natalensis* Stapf ex Hiern, Cat. Afr. Pl. Welw. **1**, 3: 665 (1898). Type from Angola.
  Rauvolfia leucopoda K. Schum. ex De Wild. & T. Durand in Bull. Soc. Bot. Belg. **38**, 1: 205 (1899). Type from Cameroon.
  Rauvolfia welwitschii Stapf, tom. cit.: 110 (1902). Type from Angola.
  Rauvolfia obliquinervis Stapf, tom. cit.: 112 (1902). Type from Tanzania.
  Rauvolfia goetzei Stapf, tom. cit.: 113 (1902). Type from Tanzania.
  Rauvolfia oxyphylla Stapf in Kew Bull. **1908**: 9: 407 (1908). Syntypes from Uganda.
  Rauvolfia tchibangensis Pellegrin in Bull. Mus. Nation. Hist. Nat. Paris **31**: 466 (1925); in Mém. Soc. Linn. Normand., N.S. **1**: 28, pl. 6 (1928). Type from Gabon.
  Rauvolfia mayombensis Pellegrin, tom. cit.: 27 (1928). Type from Congo (Brazzaville).

Tree up to 40 m. high. Trunk 5–100 cm. in diam. Bark smooth, lenticellate or rough and corky, longitudinally fissured, grey to brown; inner bark pale buff, granular; wood soft. Branches and branchlets brown, often 4–5-angular or -winged, with conspicuous leaf scars, lenticellate. Leaves usually confined to the apices of the branchlets; petiole 0–60 mm. long, glabrous; lamina 1·8–50(70) × 0·8–15(19) cm., narrowly obovate to narrowly elliptic, subacute to acute at the apex, cuneate at the base, subcoriaceous to coriaceous, glossy glabrous, with 12–35(45) secondary veins on each side of the midrib. Inflorescences many-flowered, congested in ultimate branchings; peduncle 1·5–13·5 cm. long, glabrous. Pedicels 0·1–2 mm. long, glabrous. Sepals 0·3–1·3 mm. long, triangular, glabrous. Corolla hypocrateriform; tube white, greenish- or yellowish-white, 3–5·5 mm. long, inside pubescent from 1·5–4·5 mm. from the base to the lobes; lobes sometimes different from the tube in colour, white, greenish- or yellowish-white, 0·6–1·6 mm. long, ovate to obovate, inside pubescent at the base. Stamens inserted 1·9–4·2 mm. above the base of the corolla tube; filaments 0·3–0·7 mm. long; anthers 0·6–1·2 mm. long. Pistil 1·2–3·8 mm. long; carpels 0·8–1·2 mm. long, glabrous; disk ring-shaped, 0·3–0·7 mm. high, glabrous; style 0·6–3·2 mm. long, glabrous; clavuncula 0·3–0·6 mm. long, often sparsely pubescent; stigma 0·1–0·2 mm. long, glabrous. Fruits dark red, often lenticellate; when apocarpous only one carpel developing (sub-globose and 5–20 mm. in diam.); when syncarpous bilobed, 10–30 × 7–23 × 7–24 mm. Seeds 7–13 × 4–10 × 2–4 mm.

Zambia. B: Kasisi R., c. 1050 m., fl. 5.ix.1953, *Gilges* 257 (K; PRE; SRGH). N: near Mbala (Abercorn), fr. 18.xi.1952, *White* 3714 (BR). W: Ndola, fr. 29.ix.1954, *Fanshawe* 1581 (BR; K). C: Chakwenga Headwaters, fr. 28.x.1963, *Robinson* 5788 (K; M; SRGH). E: E. of Machinje Hills, 900 m., fl. & fr. immat. 13.x.1958, *Robson* 103 (BM; BR; K; LISC; PRE; SRGH). S: R.

Tab. 108. RAUVOLFIA CAFFRA. 1, habit ($\times\frac{2}{3}$), from *Leeuwenberg* 10810; 2, bark ($\times\frac{2}{3}$), from *Leeuwenberg* 12391; 3, flower ($\times6$); 4, part of corolla opened out ($\times6$); 5, stamen, adaxial side ($\times14$); 6, stamen, lateral view ($\times14$), 3–6 from *Leeuwenberg* 10868; 7, 8, fruits ($\times\frac{2}{3}$); 9, seed ($\times\frac{2}{3}$); 10, embryo ($\times\frac{2}{3}$), 7–10 from *Leeuwenberg* 10873.

Kafue, fl. & fr. 6.x.1957, *Angus* 1750 (BR; K; LISC; P; PRE; SRGH). **Zimbabwe.** N: Mazoe Distr., Chipoli, c. 840 m., fl. 13.ix.1958, *Moubray* 6 (SRGH). W: Nyamandhlovu, c. 1200 m., fl. & fr. vii.1911, *Mundy* 1081 (SRGH). C: Shurugwi (Selukwe) Distr., c. 900 m., fr. 8.xii.1953, *Wild* 4303 (K; LISC; MO; PRE; SRGH). E: Mutare (Umtali) Distr., Park River Mutare, c. 1065 m., fl. 26.v.1961, *Chase* 7497 (COI; K; LISC; M; MO; NY; P; PRE; UPS). S: Zimbabwe Distr., c. 1050 m., fl. 4.x.1949, *Wild* 2993 (K; LISC; PRE; SRGH). **Malawi.** N: Mzimba Distr., 4 km. S.W. of Chikangawa, c. 1710 m., fl. ix.1978, *Phillips* 3884 (K; WAG). C: Ntchisi (Nchisi) Mt., 1500 m., fl. 11.ix.1946, *Brass* 17617 (A; BM; BR; K; MO; NY; PRE; SRGH; UC; US). S: Thyolo Mt., 1200 m., buds. 24.ix.1946, *Brass* 17773 (K; MO; NY; SRGH; US). **Mozambique.** N: Between Ribáuè and Iapala, fr. 16.x.1948, *Pedro & Pedrogão* 5523 (LMU). Z: Quelimane Distr., Mocuba, Mungulini (Mguluni) Mission, fl. ix, *Faulkner* 58 (BR; K; LISJC; S; SRGH). T: Macanga, near Furancungo, fl. 29.ix.1942, *Mendonça* 496A (LISC). MS: Mossurize, Serra de Espungabera (Spungabera), c. 1030 m., fr. galls. 15.iii.1966, *Pereira, Sarmento & Marques* 1399 (BR; LMU; WAG). M: Marracuene (Maracuene), Bobole, fr. 3.x.1957, *Barbosa & Lemos* 7945 (COI; K; LISC).

Widely distributed in tropical Africa from Ghana to Angola in the west and furthermore from Sudan to S. Africa. In rain-forest, riverine forests and old secondary forest.

4. **Rauvolfia mombasiana** Stapf in Kew Bull. **1894**: 21 (1894); in F.T.A. **4**, 1: 114 (1902). Syntypes from Tanzania, Kenya and Mozambique: Lower Zambezi, Shupanga, 10.i.1863, *Kirk* s.n. (K).

> *Rauvolfia monopyrena* K. Schum. in Abh. Preuss. Akad. Wiss. **1894**: 27 (1894); in Engl., Pflanzenw. Ost-Afr. **C**: 318 (1895). Type from Tanzania.

Shrub or small tree, 0·50–9 m. high. Trunk up to 10 cm. in diam. Bark grey-brown, smooth or rough, lenticellate. Branches medium-green. Leaves petiolate; petiole 5–32 mm. long, glabrous; lamina 2·5–23·5 × 1–7 cm., narrowly obovate to narrowly elliptic, acute to subacuminate at the apex, cuneate at the base, subcoriaceous to coriaceous, glossy, glabrous, with 8–25 secondary veins on each side of the midrib. Inflorescences many-flowered, fairly lax. Peduncle 2·2–15·5 cm. long, glabrous. Pedicels 3–8 mm. long, glabrous. Sepals 0·5–1·3 mm. long, triangular, glabrous. Corolla hypocrateriform; tube white, yellowish- or greenish-white, 4–7·6 mm. long, inside pubescent from 2·8–5·8 mm. above the base (or almost) to the apex; lobes sometimes different from the tube in colour, white, yellowish- or greenish-white, 0·8–1·8 mm. long, ovate to obovate, glabrous. Stamens inserted 3–6·8 mm. above the base of the corolla-tube; filaments 0·2–0·7 mm. long; anthers 0·8–1·2 mm. long. Pistil 2·6–5·5 mm. long; carpels 1·1–1·7 mm. long, glabrous; disk ring-shaped, 0·4–1 mm. high, glabrous; style 2–4·6 mm. long, glabrous; clavuncula 0·6–1 mm. long, sometimes sparsely pubescent; stigma 0·1–0·2 mm. long, glabrous. Fruits orange, yellowish- or reddish-orange; one or both carpels developing; carpels ellipsoid, 5–10 × 4–8 mm. Seeds 3–8 × 2–6 × 2–5 mm.

Mozambique. N: Eráti, km. 10 Namapa-Nacaroa, c. 300 m., fl. 11.xii.1963, *Torre & Paiva* 9508 (LISC). Z: Namacurra, km. 45 Nicuadala-Campo, c. 40 m., fr. 2.ii.1966, *Torre & Correia* 14377 (LISC). MS: Lower Zambezi, Shupanga, fl. 10.i.1863, *Kirk* s.n. (K).

Also in Kenya and Tanzania. Sandy or rocky places near the coast, on riverbanks, in light forest or in woodland.

20. PACHYPODIUM Lindley

Pachypodium Lindley, Bot. Reg.: t. 1321 (1830). — Pichon in Mém. Inst. Sci. Madag., Sér. B, **2**: 98 (1949).

> *Belonites* E. Mey., Comm. Pl. Afr. Austr.: 187 (1837).

Shrubs or small trees with thick succulent trunk and stems. Latex present. Conspicuous ternate spines (the central one often very reduced or absent) present at each node. Leaves alternate or spirally arranged. Cymes terminal, sessile or pedunculate, few–many-flowered; flowers usually showy, pink, purple, yellow or partly white. Calyx lobes imbricate, free almost to the base. Corolla hypocrateriform to subcampanulate, the tube constricted near the base, tubular to campanulate above; lobes contorted, overlapping to the right. Stamens inserted above the constriction; anthers subsessile, conniving in a cone, the fertile thecae confined to the upper part of the anther below which a brush-like retinacle projects inwards from the connective, adhering to the clavuncle. Disc cupular, lobed or replaced by 5 scale-like processes. Ovary of 2 free carpels each containing ∞ ovules on the ventral placenta;

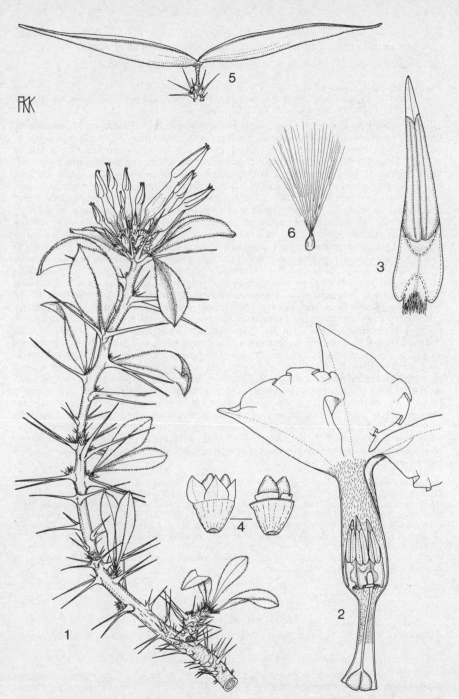

Tab. 109. PACHYPODIUM SAUNDERSII. 1, habit (× ⅔), from *Leach* 10801; 2, part of corolla, opened out (× 2); 3, stamen, ventral view (× 6); 4, (left) thickened pedicel surmounted by calyx, with the two free carpels within; (right) the same with sepals removed to show the disc-scales (× 2), 2–4 from *Sousa* 196; 5, fruit (× ⅔), from *Chase* 2247; 6, seed (× ⅔), from *Correia & Marques* 1196.

style slender, cylindric; clavuncle annular, style 2-lobed. Fruit of 2 follicles. Seeds ∞, ovate to oblong with an apical coma, endosperm scanty or absent.

A genus of 18 species, 13 native to Madagascar, the rest to southern Africa.

Pachypodium saundersii N.E. Br. in Kew Bull. **1892**: 126 (1892). — Stapf in F.C. **4**. 1: 516 (1907). — Phillips in Fl. Pl. S. Afr. **4**: t. 123 (1924). — Codd in Fl. Southern Afr. **26**: 286 (1963). TAB. **109**. Type from S. Africa (Natal).

Low succulent shrub 0·5–2 m. high with swollen, little-branched stems arising from a large tuber. Stems glabrous, bark smooth, pale grey, deeply longitudinally wrinkled on drying. Leaves spirally arranged, each lamina subtended by a pair of spines, these 10–40 mm. long, stout, glabrous, straight, angled slightly upwards, with confluent swollen bases; a third, much smaller, spine also usually present proximal to the petiole. Leaves 25–78 × 10–25 mm., thin-textured, drying blackish, often ephemeral; petiole 0–3 mm.; lamina narrowly obovate with apex obtuse and mucronate or cuspidate-acute, base tapering; upper surface glabrous, midrib ± level, lateral veins obscure; margin usually tuberculate-hispid; lower surface completely glabrous or with hispid midrib, the latter raised, other nerves usually obscure; leaf sometimes strongly undulate so that it cannot be pressed flat without folding. Short shoots, with crowded slender spines up to 1 cm. long, present in axils of main stem leaves. Inflorescences terminal, many-flowered, very condensed; axes glabrous; flowers pink to purple on outside of corolla, white above, greenish within. Calyx c. 4 mm. long, glabrous, segments imbricate, triangular. Corolla tube 30–42 mm. long, narrow in the lowest third then widening abruptly and gradually narrowing towards the apex, externally glabrous, internally pilose with long weak hairs above and below the stamens. Corolla lobes $\frac{1}{2}$–$\frac{2}{3}$ as long as tube, asymmetrical, ± triangular with one straight and one strongly rounded, crisped edge; both surfaces glabrous. Anthers sessile, inserted at base of swollen part of corolla tube, 9·5–11 mm. long, each comprising a sterile sagittate structure bearing two thecae 6–6·5 mm. long on its ventral surface and a retinacle. Five partially united oppositipetalous glabrous scales c. 1 mm. long borne on receptacle internal to the corolla. Ovary of 2 completely free, compressed-ovoid, glabrous carpels, each unilocular with ventral placenta bearing ∞ (c. 200) ovules. Style 1, slender, glabrous, attached to the two carpels at their apices; clavuncle annular, joined to the anther retinacles by a rubbery latex; style 2-lobed. Fruit of 2 horn-like follicles up to 15 cm. long containing ∞ seeds. Seeds c. 7 mm. long, compressed-ovoid, with apical coma of pale golden hairs up to 5 cm. long.

Zimbabwe. E: Chipinge Distr., Sabi Valley, near Rimayi, fl. 11.iii.1965, *Plowes* 2685 (LISC; PRE; SRGH). S: Nuanetsi Distr., 3 km. S. of Lundi R. bridge on Masvingo-Beitbridge road, fl. 21.iv.1961, *Leach* 10801 (BM; COI; PRE; SRGH). **Mozambique.** MS: Maringuè, 10 km. N. of Sabi R., fr. 23.vi.1950, *Chase* 2247 (BM; SRGH). GI: Gaza, between Caniçado and Mapai, fl. 5.v.1944, *Torre* 6579 (C; LISC; LMA; MO). M: Goba, Pioneers' Memorial (Monumento dos Pioneiros), fl. & fr. 10.vi.1945, *Esteves de Sousa* 196 (K; LISC; LMU; WAG).

Also occurring in S. Africa (Natal and northern and eastern Transvaal) and Swaziland. In open woodland among rocks in full sun.

21. ADENIUM Roem. & Schult.

Adenium Roem. & Schult.; Syst. Veg. **4**: 35 (1819). — Plaizier in Meded. Landb. Wag. **80**–12: 3 (1980).
Idaneum Post & Kuntze, Lex. Phan.: 296 (1904).

Succulent shrubs or trees with rhizomatous or carrot-like tubers; with clear or white latex. Leaves alternate, confined to the apices of the branchlets, with colleters in the axils; stipules minute or absent. Inflorescence thyrsoid, lax; bracts narrowly obovate to linear, acuminate, entire, pubescent; peduncle very short or absent; pedicels pubescent. Flowers slightly zygomorphic. Sepals connate at the base, subequal, narrowly oblong to narrowly obovate, entire, acuminate. Corolla tube infundibuliform to hypocrateriform, much widened at the throat, more or less pubescent outside, glabrous or pubescent to strigose inside; lobes obovate or narrowly obovate, acuminate, entire, undulate or crispate, spreading, overlapping to the right in bud; between all the lobes an obcordate, glabrous or pubescent to velutinous scale at the base, which is united by its edges with the lobes. Stamens included to exserted, inserted at the apex of the narrow basal portion of the corolla-tube; anthers narrowly triangular, sagittate at the base, and with a long apical appendage, appendages filiform, usually coherent and twisted at the apex. Carpels two, globose, connate at the base; style split only at the base, cylindrical; clavuncula subcylindrical, more or less coherent with the apices of the filaments; stigma bifid, c. 0·5 mm. long. Fruit consisting of two spreading or recurved follicles; follicles connate at the very base, oblong and tapering towards each end, pubescent outside, glabrous inside, many-seeded. Seeds oblong, truncate, with tufts of dirty white to light brown hairs at both ends.

A genus of 5 species in the tropics of South Arabia, western, central, northeastern and southern Africa and in Socotra.

1. Leaves linear, very narrowly obovate or narrowly oblong; secondary veins inconspicuous; flowers usually appearing with the leaves - - - - - - - - 2
 - Leaves obovate, ovate, oblong or less often narrowly obovate; secondary veins conspicuous; flowers appearing with or often before the leaves - - - - - - - 3
2. Leaves obtuse to rounded at the apex, shortly petiolate; petiole 1–4 mm. long; stamens included; corolla tube glabrous inside, narrow basal portion 0·6–1 times as long as the calyx
 3. *swazicum*
 - Leaves acute, sessile; stamens barely included or exserted; corolla-tube puberulous inside, narrow basal portion (1)1·5–2 times as long as the calyx - - - - 2. *oleifolium*
3. Narrow basal portion of the corolla-tube 0·6–1 times as long as the calyx; stamens included; corolla-tube glabrous inside; lobes undulate; leaves pubescent at least beneath 3. *swazicum*
 - Narrow basal portion of the corolla-tube 1·1–3·3(3·8) times as long as the calyx; stamens distinctly exserted; corolla-tube pubescent inside; lobes crispate; leaves glabrous
 1. *multiflorum*

1. **Adenium multiflorum** Klotzsch Peters, Reise Mosamb., Bot.: 279, t. 44 (1861). — Stapf in F.T.A. **4**, 1: 229 (1902). — Codd in Fl. Southern Afr. **26**: 279 (1963). — Plaizier in Meded. Landb. Wag. **80**–12: 9, fig. 2, phot. 1, map 2 (1980). TAB. **110**. Type: Mozambique, near Tete, *Peters* s n (B†, holotype); between Mopeia and Campo, *Mendonça* 2040 (LISC, neotype, designated by Plaizier, loc. cit.).
  Adenium obesum var. *multiflorum* (Klotzsch) Codd in Bothalia **7**: 452 (1961). Type as above.
  Adenium multiflorum subsp. *multiflorum* (Klotzsch) Rowley in Cactus and Succ. Journ. (U.S.) **46**: 164 (1974). Type as above.

Succulent shrubby tree, 0·5–3·5 m. high with a large carrot-like root, up to 1 m. in diameter with poisonous white latex; bark shiny, grey. Leaves usually appearing after the flowers, subsessile; lamina obovate to oblong, 1·5–3(5·5) times as long as wide, (3·5)7·6–12·5 × (1·4)2–7·6 cm., acute or rounded to emarginate and apiculate to mucronate at the apex, glabrous on both sides; secondary veins conspicuous, (5)6–11(13); petiole 3–7 mm. long. Inflorescence 0·75–2 × 0·5–1·5 cm., bracts narrowly obovate, 4–6 × 1–3 mm., acuminate. Pedicels densely pubescent to tomentose, 2–4 mm. long. Sepals narrowly ovate, 6–10 × 2·5–3 mm., pubescent outside, appressed-pubescent inside, especially towards the apex. Corolla red to white; tube pink to white, red-striped within the throat (2)2·5–4 times as long as the calyx, 2·2–3·9 × 1–1·3 cm. pubescent outside, sometimes towards the extreme base somewhat less pubescent inside with a pubescence of obscure glandular hairs, usually on the main veins with strigose glandular hairs; narrow basal portion 0·9–1·1(1·4) times as long as the calyx, 7–10 × 3–7 mm.; lobes pink to white with deep pink to scarlet margins, 1–3–2·9 × 1–1·9 cm., narrowly ovate to narrowly obovate, mucronate to apiculate, crispate, outside sparsely and minutely pubescent, glabrous inside, with a velutinous scale, 2 × 2–2·5 mm., at the base. Stamens distinctly exserted; free portion of filament densely pubescent outside, lanate inside; anther 5·5–7 × 0·5–1 mm.; thecae 2–3 × 0·5 mm.; appendages pink to white, 2–4·9 times as long as the anther, hispid especially on the outside. Pistil 10·5–12·5(15) mm. long; ovary glabrous; carpels 1–2 × 0·5–1 × 0·5–1·5 mm.; style 8–11 × 0·5 mm.; clavuncula 1–1·5 × 0·5–1 mm. Fruit pale grey to pale grey-brown, (7–)10–18 × 0·8–1·5 cm. Seed very pale brown, glabrous or very minutely appressed pubescent, 1–1·5 × 0·2–0·3 cm.; comas dirty white to light brown, 2–3 cm. long.

Zambia. C: Katondwe Mission, 4.ii.1964, *Fanshawe* 8247 (K). **Zimbabwe.** W: Hwange (Wankie) Distr., between Sebungwe R. and Zambezi R., fl. 14.v.1955, *Plowes* 1848 (K;

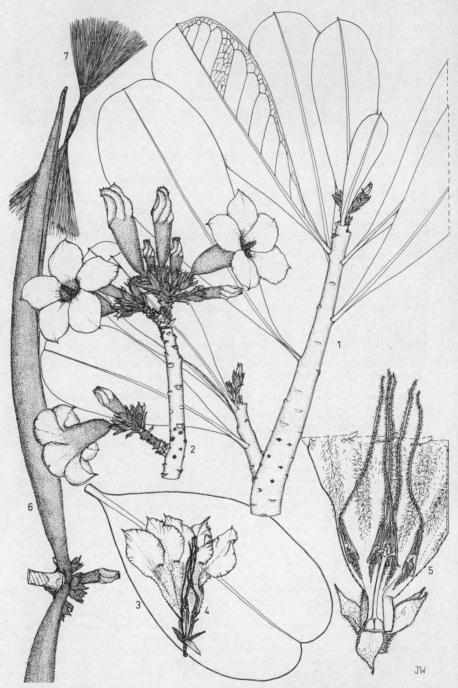

Tab. 110. ADENIUM MULTIFLORUM. 1, branch leaves (×⅔), from *Leach* 9923; 2, flowering branch (×⅔), from *Mendonça* 2040; 3, leaf (×⅔), from *Rogers* 20717; 4, flower (×⅔), from *Rodin* 4721; 5, anthers and pistil (×2), from *Taylor* 638; 6, fruit (×⅔), from *Mendonça* 2040; 7, seed (×⅔), from *Angus* 3335.

PRE). C: Arcturus Distr., Ewanrigg Nat. Park, 12.xi.1947, *Hopkins* B 1603 (UC). E: Chimanimani (Melsetter) Distr., Nyanyadzi, 13.viii.1970, *Plowes* 3433 (K; PRE). S: Masvingo (Fort Victoria) Distr., x.1920, *Eyles* 2760 (K; SRGH). **Malawi.** S: Nsanje (Port Herald) Distr., Matope-Mwabyi Game Reserve, fl. 6.viii.1975, *Salubeni* 1967 (SRGH). **Mozambique.** Z: Morrumbala, Posto do Chire, 4.iv.1972, *Bowbrick* 5208 (LISC; SRGH). T: 1·6 km. N.E. of Tete ferry, 17.viii.1960, *Leach* 10460 (K; SRGH). MS: Sofala Prov., between Pungoé & Búzi R., S. of Tica, 19.vi.1961, *Leach & Wild* 11114 (K; LISC; PRE; SRGH). GI: Inhambane Prov., Vilanculos, between Funhalouro and Saúte, 19.v.1941, *Torre* 2699 (C; COI; LISC; LMU; WAG). M: Umbelúzi, fl. 11.viii.1920, *Borle* 543 (K; PRE; SRGH).

Also in S. Africa (Transvaal, Natal) and Swaziland. At low altitudes in hot dry areas in sandy or rocky habitats, in open deciduous woodland. Alt.: 0–700 (1·200) m. Used as fish, arrow-, and magic poison. Poisonous to cattle, but probably not eaten by them.

2. **Adenium oleifolium** Stapf in Kew Bull. **1907**: 53 (1907); in F.C. **4**, 1: 513 (1907). — Codd in Fl. Southern Afr. **26**: 281 (1963). — Plaizier in Meded. Landb. Wag. **80**–12: 19, fig. 4, map 4 (1980). Type from S. Africa.
  Adenium lugardii N.E. Br. in Kew Bull. **1909**: 119 (1909). Type: Botswana, Palapye, *Lugard* 269 (K, holotype).
  Adenium oleifolium var. *angustifolium* Phillips in Fl. Pl. S. Afr. **3**: t. 105 (1923). Type from S. Africa.
  Adenium somalense var. *angustifolium* (Phillips) Rowley in Cactus & Succ. Journ. (U.S.) **46**: 164 (1974). Type as above.

Succulent shrublet up to 40 cm. high, forming a dense mass of rather fleshy leaves and stems, with a subterraneous carrot-like very bitter-tasting rootstock 50–80 × 15–30 cm. Leaves sessile; lamina linear to very narrowly obovate, 8·8–16·7(21·2) times as long as wide, 4·5–14·6 × 0·3–1·4 cm., acute and apiculate, above shiny, glaucous or pale green, and pubescent to glabrous, beneath dull, slightly paler green and pubescent. Inflorescence 0·5–1 × 0·5–1 cm.; bracts narrowly oblong, 3–4 × 1–1·5 mm. Pedicels densely pubescent to pilose, 5–8 mm. long. Sepals narrowly oblong to narrowly ovate, 6·5–9(12) × 3·0–4·0 mm., outside densely pubescent to pilose pubescent inside. Corolla bright scarlet or red to pink; tube yellowish especially towards the base, 5–6·9(8·2) times as long as the calyx, 4–6·6 × 0·9–1·4 cm., pubescent outside, only at the extreme base nearly glabrous, puberulous inside, somewhat velutinous on the main veins; narrow basal portion (1)1·5–2 times as long as the calyx, (0·8)1·2–1·7 × 0·3–0·5 cm.; lobes bright scarlet to red, obovate, 1·4–2·8 × 0·8–1·8 cm., apiculate and undulate, sparsely and minutely pubescent outside, puberulous inside, with a puberulous scale at the base, 3 × 1·5 cm. Stamens barely included or slightly exserted, free portion of filament hispid outside, pilose to lanate inside; anther 5·5–6(8) × 1–1·5 mm., hispid outside, cells 2–3·5 × 1 mm.; appendages 2·5–3·1 times as long as the anthers, hispid. Pistil 13·5–21·5 mm. long; ovary glabrous or sometimes with appressed stiff hairs or puberulous; carpels 1–2 × 1–1·5 × 1–2 mm.; style 11·5–19 × 0·5 mm.; clavuncula 1·5–2 × 1–1·5 mm. Fruit pale grey to pale grey-brown, 10–11·5 × 1 cm. Seed very pale brown, pubescent, 1–1·5 × 0·2 cm., with tufts of dirty white hairs 2–3·5 cm.

Botswana. SE: 22 km. S. of Artesia, 18.i.1960, *Leach & Noel* 237 (K; PRE; SRGH).
Also in S. Africa (Transvaal and Cape Prov.) and Namibia, on loose white or red sandy or sometimes rocky soil in bushland, 700–1200 m.

3. **Adenium swazicum** Stapf in Kew Bull. **1907**: 53 (1907); in F.C. **4**, 1: 513 (1907). — Codd in Fl. Southern Afr. **26**: 281 (1963). — Plaizier in Meded. Landb. Wag. **80**–12: 22, fig. 5, phot. 2, map 5 (1980). Type from Swaziland.
  Adenium boehmianum var. *swazicum* (Stapf) Rowley in Cactus & Succ. Journ. (U.S.) **46**: 164 (1974). Type as above.

Succulent shrub, 0·2–0·7 m. tall with a carrot-like tuber up to 1 m. in diam. with poisonous clear latex. Leaves petiolate; petiole 1–4 mm. long; lamina oblong to narrowly oblong, 3·5–9·1 times as long as wide, 4–11·5 × 0·5–3·1 cm., rounded and apiculate to mucronate, rarely emarginate, above pubescent, especially the midrib; secondary veins more or less inconspicuous, beneath pubescent. Inflorescence 1·5–3·5 × 1–2·5 cm.; bracts narrowly oblong to narrowly ovate, 3–10 × 2 mm. Pedicels 6–10(15) mm. long, tinged with pink or red. Calyx crimson or pink to green, narrowly oblong to narrowly ovate, 7–11 × 1·5–3 mm., pubescent outside, appressed-pubescent inside, especially towards the apex. Corolla crimson, deep mauve or pink to white; tube crimson to white, 2·2–3·5(4) times as long as the calyx, 2–3 × (0·6)1–1·3(1·9) cm., outside pubescent, only at the very base nearly glabrous, glabrescent inside; narrow basal portion 0·6–1 times as long as the calyx, 0·5–0·9 × 0·2–0·4 cm.; lobes deep mauve to white, obovate, 1·3–2·5(3·5) × 1–2 cm., apiculate, slightly undulate, both sides puberulous; a glabrous scale at the base (2)2·5 × 1·5 mm. Stamens included; free part of filament glabrous outside, lanate inside; anther 5–6·5 × 1–1·5 mm., hispid outside; cells 2–3 × 1 mm.; appendages (1·2)1·5–2 times as long as the anther, hispid. Pistil 9–11·5 mm. long; ovary glabrous or puberulous to sericeous; carpels 1·5–2·5 × 1–1·5 × 1–2·5 mm.; style 5·5–7(9·5) × 0·5 mm.; clavuncula 1–1·5 × 0·5–1 mm. Follicle grey-brown, 16 × 1 cm. Seed pale-brown, glabrous, 1·2–1·4 × 0·3 cm., comas dirty white, 2·8–3·5 cm. long.

Mozambique. M: between Magude and Chobela, fl. 21.i.1944, *Torre* 6375 (LISC; PRE).
Also in S. Africa (eastern Transvaal and northern Zululand) and Swaziland. In open woodland on sand and often brackish soil, 300–400 m.

22. STROPHANTHUS DC.

Strophanthus DC. in Bull. Soc. Philom. **64**: 122 (1802). — Beentje in Meded.
Landb. Wag. **82**–4: 17 (1982).
Cercocoma Wall. ex G. Don, Gen. Syst. **4**: 84 (1837).
Christya Ward & Harvey in Journ. Bot. Lond. **4**: 134 (1842).
Roupellia Wall. & Hook. in Bot. Mag. **75**: t. 4466 (1849).
Zygonerion Baill. in Bull. Mens. Soc. Linn. Paris **1**: 758 (1888).

Lianas or sarmentose shrubs; latex present. Trunk dichotomously or trichotomously branched or with single or opposite lateral branches, unarmed, lenticellate. Leaves decussate or ternate, rarely quaternate, those of a pair or a whorl equal or subequal, with 2–20 colleters in the leaf axil; lamina ovate, elliptic or obovate. Inflorescence terminal (often on short lateral branches) or in the forks of branches, one to many-flowered in cymes; bracts sepal-like or smaller in size, in the axil with colleters. Flowers actinomorphic or with only the sepals unequal. Sepals imbricate, free to the base, at the base inside with colleters. Corolla consisting of tube, corona, and lobes; colour scheme based on contrasting white and red, later turning yellow and purple; tube widening at the level where the anthers are inserted; corona 10-lobed with the lobes arranged in partly connate pairs at the basal margins of the lobes; lobes often produced into long, linear tails overlapping to the right, in open flowers spreading or recurved, basal part of lobe ovate. Stamens included or partly exserted, anthers connivent in a cone around the apex of the style, auriculate at the base, mucronate or acuminate at the apex. Disk absent. Ovary 2-celled, semi-inferior or rarely superior, ovoid, with many ovules; style slender, glabrous; clavuncula with a basal frill and an apical crown surrounding the small stigma. Fruit of 2 divergent, rigid, woody or thin-woody follicles, many- or rarely few-seeded, tapering to their apex and ending in a narrow point or a knob. Seeds rostrate, with a rostrate apical coma; grain almost fusiform; rostrum often partly glabrous; endosperm in a thin layer completely surrounding the spathulate, straight embryo.

A genus of 38 species distributed in Asia, Madagascar and Africa, with 30 species in Africa.

Key to flowering material

1. Flowers and leaves simultaneously present - - - - - - - - - 2
– Flowers on plants in the leafless stage - - - - - - - - - - 14
2. Corolla lobes acute or acuminate, but not tailed - - - - - - - 3
– Corola lobes produced into long, linear tails - - - - - - - - 5
3. Anthers exserted for 8–12 mm., with an acumen of 17·5–20 mm. long; narrow part of tube 13–19 mm. long; corona lobes long-pubescent - - - - - 3. *gardeniiflorus*
– Anthers included for 2–16 mm., with an acumen of less than 2 mm. long; narrow part of tube 3–12 mm. long; corona lobes puberulous - - - - - - - 4
4. Branchlets glabrous; stems and older branches with corky protuberances; corona lobes obtuse, 2–6 mm. long; petiole 3–11 mm. long - - - - - 1. *courmontii*
– Branchlets minutely puberulous; no corky protuberances present; corona lobes acute, 5–23 mm. long; petiole 1–5 mm. long - - - - - 12. *welwitschii*
5. Leaves in whorls of 3 (sometimes a few branches with leaves opposite or in whorls of 4 also present) - - - - - - - - - - - - - - 11. *speciosus*
– Leaves opposite- - - - - - - - - - - - - - - 6

6. Branchlets glabrous; stems and older branches often with corky protuberances - 7
 - Branchlets puberulous, pubescent or hispid, without corky protuberances - - 8
7. Leaves 1–2 cm. wide, with an acumen of less than 3 mm. long and inconspicuous tertiary venation; corona lobes 2–5 mm. long; corolla lobes including the tails 30–59 mm. long
4. gerrardii

 - Leaves 17–52 cm. wide, with an acumen of 2–10 mm. long and conspicuous tertiary venation; corona lobes 6–15 mm. long; corolla lobes including the tails 90–205 mm. long
10. petersianus

8. Corolla tube glabrous outside - - - - - - - - 10. <i>petersianus</i>
 - Corolla tube puberulous or tomentellous outside - - - - - - - 9
9. Branchlets, leaves and inflorescence hispid with 1–2 mm. long stiff hairs- - 7. <i>kombe</i>
 - Branchlets, leaves and inflorescence puberulous or pubescent (leaves sometimes glabrous in <i>S. luteolus</i>) with supple hairs less than 1 mm. long - - - - - 10
10. Leaves green beneath; filaments curved, 0·9–1 mm. high; anthers puberulous near the filament only - - - - - - - - - - 8. <i>luteolus</i>
 - Leaves yellowish or silvery beneath because of the indument; filaments straight, 1–6·2 mm. long; anthers glabrous - - - - - - - - 11
11. Mature leaves more than 10 cm. long; calyx 10–25 mm. long; corolla tube 17–26 mm. long, lobes (including the tails) 94–180 mm. long - - - - - - 12
 - Mature leaves less than 7·5 cm. long; calyx 4·5–12·5 mm. long; corolla tube 10–20 mm. long, lobes (including the tails) 22–115 mm. long - - - - - 13
12. Shrub or tree, rarely lianescent; corona lobes 2·6–6·5 mm. long; calyx usually eglandular, rarely with 10 colleters; petiole 1–10 mm. long; branches often semi-succulent 2. <i>eminii</i>
 - Liana; corona lobes 1·5–3 mm. long; calyx with 5 colleters; petiole 1–5 mm. long; branches not succulent - - - - - - 5. <i>holosericeus</i>
13. Corolla lobes (including the tails) 22–57 mm. long; bracts 4–16 × 2·5–7 mm.; sepals 2–9 mm. wide - - - - - - - - - 6. <i>hypoleucos</i>
 - Corolla lobes (including the tails) 55–115 m. long; bracts 2–5 × 1–2 mm.; sepals 1·3–3 mm. wide- - - - - - - - - - 9. <i>nicholsonii</i>
14. Corolla tube widening below the middle of the tube; ovary glabrous; anther-acumen 1–4 mm. long; stems and older branches often with corky protuberances - - 15
 - Corolla tube widening at or above the middle of the tube; ovary pubescent or hispid; anther-acumen less than 0·6 mm. long; corky protuberances absent - - 16
15. Corona lobes 2–5 mm. long; corolla lobes (including the tails) 30–59 mm. long
4. gerrardii

 - Corona lobes 6–15 mm. long; corolla lobes including the tails 90–205 mm. long
10. petersianus

16. Sepals 4·5–12 mm. long; anthers 4–5·3 mm. long; corolla tube 10–22 mm. long; corolla-lobes less than 60 mm. long, or, if 60–115 mm. long, the style 6–10 mm. long - - 17
 - Sepals 8–25 mm. long; anthers 5–7 mm. long; corolla tube 17–26 mm. long; corolla lobes 94–200 mm. long, style 11–18·5 mm. long - - - - - - 12
17. Corolla lobes 22–57 mm. long; bracts 4–16 × 2·5–7 mm.; sepals 2–9 mm. wide
6. hypoleucos

 - Corolla lobes 55–115 mm. long; bracts 2–5 × 1–2 mm.; sepals 1·3–3 mm. wide
9. nicholsonii

Key to fruiting material

1. Follicles shaggy with pubescent protuberances - - - - - 2. <i>eminii</i>
 - Follicles smooth or with only a few protuberances - - - - - 2
2. Leaves ternate (sometimes a few branches with opposite or quaternate leaves also present); seed rostrum glabrous for 0–2 mm. - - - - - 11. <i>speciosus</i>
 - Leaves opposite; seed rostrum glabrous for more than 4 mm. - - 3
3. Branchlets glabrous - - - - - - - - - - - - 4
 - Branchlets puberulous, pubescent or hispid - - - - - - 6
4. Stem and older branches with corky protuberances; seed rostrum glabrous for more than 13 mm. - - - - - - - - - - - - - 5
 - Corky protuberances absent; seed rostrum glabrous for 4–11 mm.- - 3. <i>gardeniiflorus</i>
5. Leaves 3–6 × 1–2 cm., with inconspicuous tertiary venation and an acumen of less than 3 mm.; fruit 1·5–2 cm. in diam. - - - - - - 4. <i>gerrardii</i>
 - Leaves 2·5–13 × 1·7–6·5 cm., with conspicuous tertiary venation and an acumen of 1–13 mm. long; fruit in diam. 2·2–4·5 cm. - - - 1. <i>S. courmontii</i> or 10. <i>S. petersianus</i>*
6. Branchlets and leaves hispid - - - - - - - 7. <i>kombe</i>
 - Branchlets and leaves (beneath) puberulous or tomentellous - - - - 7
7. Leaves tomentose or tomentellous beneath - - - - - - 8
 - Leaves puberulous beneath - - - - - - - - 10

* Virtually indistinguishable at this stage, although the lenticels on the branches are very dense in <i>S. petersianus</i> and sparse to dense in <i>S. courmontii</i>; and the lenticels on the follicles are very dense in <i>S. courmontii</i> and sparse to dense in <i>S. petersianus</i>.

8. Tertiary venation of leaves conspicuous; seed rostrum glabrous for 37–64 mm. - 5. *holosericeus*
– Tertiary venation of leaves inconspicuous; seed rostrum glabrous for 13–44 mm. - 9
9. Petiole 1–3 mm. long; seed grains 12–22 mm. long; lenticels on follicles elongate - 9. *nicholsonii*
– Petiole 2–7·5 mm. long; seed grains 8–12·5 mm. long; lenticels on follicles round or nearly so - - - - - - - - - - - - - - - - 6. *hypoleucos*
10. Petiole (2)3–13 mm. long; stems and older branches with corky protuberances - 10. *petersianus*
– Petiole 1–5 mm. long; corky protuberances absent - - - - - - - - 11
11. Liana; follicles tapering to a knob; seeds 2–3 mm. wide - - - - - 8. *luteolus*
– Shrub or liana; follicles tapering to a narrow tip or rarely a knob; seeds 2·5–4 mm. wide - 12. *welwitschii*

1. **Strophanthus courmontii** Sacl. ex Franch. in Journ. Bot., Paris **7**: 300 (1893), "*courmonti*". — Stapf in F.T.A. **4**, 1: 182 (1902). — K. Braun in Pflanzer **6**: 292 (1910). — Verdcourt & Trump, Comm. Pois. Pl. E. Afr.: 136 (1969). — Beentje in Meded. Landb. Wag. **82**–4: 62, fig. 14, map 12 (1982). TAB. **111**. Type from Tanzania.
S. courmontii var. *fallax* Holmes in Pharm. Journ. **4**, 12: 487 (1901). Type: Malawi, *Buchanan* 1219 (K, holotype; BM, E, isotypes).
S. courmontii var. *kirkii* Holmes, loc. cit. Type from Tanzania.

Liana, 5–22 m. high, or less often a sarmentose shrub, 0·60–4 m. high, deciduous with the flowers appearing after the leaves; latex white. Trunk up to 10 cm. in diam. with corky ridges up to 5 cm. long and 1·8 cm. high; branches with corky laterally compressed protuberances at the nodes, later growing into ridges, branchlets glabrous. Leaves petiolate; petiole 3–11 mm. long, lamina dark green above, much paler beneath, elliptic, ovate, or rarely obovate, 1–3·1 times as long as wide, 2·5–13·5 × 2·5–6·5 cm., rounded or cuneate at the base, mucronate or acuninate at the apex (acumen up to 10 mm. long) papyraceous or thinly coriaceous, glabrous on both sides, 3–7(8) pairs of secondary veins forming an angle of 35–60° with the midrib; tertiary venation conspicuous beneath. Inflorescence 1–3-flowered sessile or pedunculate, glabrous or sparsely puberulous in all parts; pedicels 1–7·5 mm. long; bracts ovate or narrowly ovate, 1·5–4 × 0·8–1·2 mm., acute, not sepal-like. Flowers fragrant, equal or subequal, the outer sometimes shorter and wider than the inner, ovate or narrowly ovate, 2–6 times as long as wide, 3·5–10 × 1·5–3·5 mm., acute or apiculate, glabrous, ciliate, or rarely puberulous; with 2–3 colleters per sepal, colleters rarely forked. Corolla tube white and red, purple-streaked, (22)25–43 mm. long, widening at $\frac{1}{7}$–$\frac{1}{3}$ ($\frac{2}{5}$) of its length into a cyathiform upper part, at the mouth 17–35 mm. wide, glabrous or less often puberulous outside, puberulous inside; corona lobes 2–6 × 1·2–3·2 mm., puberulous; corolla lobes ovate and gradually narrowing into the acute apex (20)25–57 × 10–27 mm., glabrous or less often puberulous on both sides, but always puberulous inside at the base. Stamens included for 5–16 mm.; filaments straight, 3·5–6 mm. long; anthers 6·5–9 × (1·3)1·6–2·2 mm., glabrous; acumen 0·8–2 mm. long. Ovary 1·3–2·2 × 2–3·8 mm., glabrous; style 8·8–15·5 mm. long; clavuncula 2·2–3 × 1·9–3 mm., stigma minute. Follicles divergent at an angle of 160–200°, tapering into a broad or narrow obtuse apex, 12–26 cm. long and 3–4·5 cm. in diameter; exocarp grey-black or purplish-black, thick and hard, glabrous, very densely lenticellate. Seeds with the grain 10–15 × 2·2–4 × 1 mm., densely pubescent; rostrum glabrous for 15–52 mm. and bearing a coma for 17–35 mm.; coma 36–70 mm. long.

Zambia. C: Feira, 5.xii.1968, *Fanshawe* 10465 (K; SRGH). E: Lutembwe R. gorge, E. of Machinje Hills, 13.x.1958, *Robson & Angus* 93 (BM; BR; K; LISC; PRE; SRGH). **Zimbabwe.** N: Zambezi R. bank, W. of Mana Pools Game Res., 26.x.1963, *West* 4575 (K; SRGH). E: Chimanimani Distr., Lusitu R. valley below Glencoe For. Res., x.1966, *Goldsmith* 85/66 (BR; K; L; LISC; M; MO; PRE; SRGH). **Malawi.** N: Likoma Isl., Lake Malawi (Nyasa), viii.1887, *Bellingham* (BM). S: Mulanje Distr., Tuchila For. Res., 1.x.1950, *Bain* 1 (K; NY). **Mozambique.** N: Amaramba, between Lúrio and Mepanhira, 19.x.1948, *Pedro & Pedrogão* 5567 (NY; PRE). M: Ribáuè area, 1948, *Andrada* 1407 (BM; COI; LISC). Z: Mocuba, Namagoa plantation, *Faulkner* in GHS 11733 (BR; K; LISJC; NY; S; SRGH). T: Mecangádzi R., Cahora Bassa, 9.xi.1973, *Correia et al.* 3768 (B). MS: Sambanke, between Mutarara de Lucite and Dombe, 13.x.1953, *Gomes Pedro* 4300 (K; LISJC; PRE). GI: Cubane, 5.i.1905, *Le Testu* 545 (BM; BR; P).

Also in Kenya and Tanzania. In riverine forest and thickets. Flowering at the end of the dry season and to some extent in the rainy season; mature fruits towards the end of the rainy season.

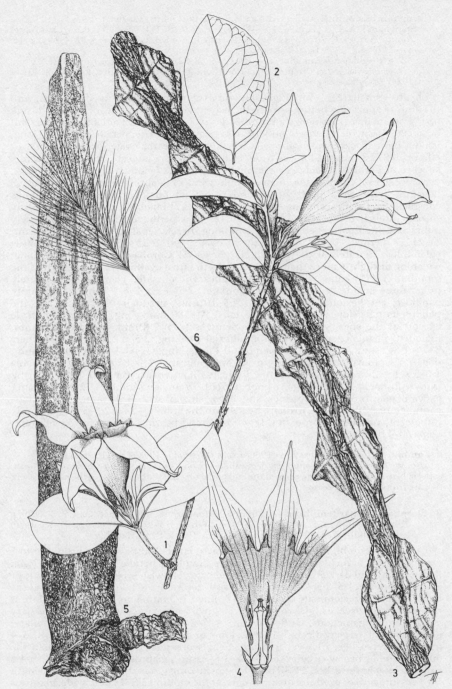

Tab. 111. STROPHANTHUS COURMONTII. 1, flowering branch ($\times \frac{2}{3}$), 2, leaf ($\times \frac{2}{3}$), both from *Helg*, Jan. 1953; 3, older branch ($\times \frac{2}{3}$), from *Harris* 2485; 4, section of flower ($\times 1$), from *Haerdi* 215/0; 5, fruit, one follicle removed ($\times \frac{2}{3}$), from *Pedro* 3319; seed ($\times \frac{2}{3}$), from *Topham* s.n.

2. **Strophanthus eminii** Aschers. & Pax in Engl., Bot. Jahrb. **15**: 366, t. 10 & 11 (1892). — Stapf in F.T.A. **4**, 1: 172 (1902). — White, F.F.N.R.: 351 (1962). — Verdcourt & Trump, Comm. Pois. Pl. E. Afr.: 37 (1969). — Beentje in Meded. Landb. Wag. **82**–4: 69, fig. 16, map 14 (1982). Type from Tanzania.

   S. wittei Staner in Ann. Soc. Sci. Brux. Sér. B, **52**: 90 (1932). Type from Zaire.

   S. eminii var. *wittei* (Staner) Staner & Michotte in Bull. Jard. Bot. État Brux. **13**: 34 (1934). Type as above.

Shrub or small tree, 1–7 m. high, sometimes climbing up to 10 m., deciduous, with the flowers appearing before or with the leaves; latex clear, white, or yellow. Trunk up to 6 cm. in diam.; branches sometimes fleshy, with adherent or loose bark, smooth or sulcate; branchlets densely puberulous. Leaves petiolate; petiole 1–10 mm. long, lamina light to dark green above, silvery-grey beneath, ovate, broadly ovate, or elliptic, 0·8–2·3 times as long as wide, 6–24 × 4–18 cm., cuneate, rounded, or rarely subcordate at the base, acute or acuminate at the apex (acumen up to 10 mm. long), papyraceous or chartaceous, densely pubescent or glabrescent above, tomentose beneath, with 5–12 pairs of secondary veins; tertiary venation sometimes conspicuous. Inflorescence on long or short leafless branches, axillary or apparently so, only rarely in the forks, 1–12-flowered, sessile or pedunculate, congested, densely pubescent in all parts; pedicels 1–8 mm. long; bracts sepal-like. Flowers fragrant. Sepals subequal, ovate or narrowly ovate, (8)11–25 × 2·5–13 mm., acute; eglandulose or rarely with 10 colleters per calyx. Corolla-tube 17–26 mm. long, widening at slightly more than ½–⅘ of its length into a cyathiform upper part, at the mouth 8–17(21) mm. wide, densely pubescent outside, glabrous or minutely puberulous inside; corona-lobes subulate, 2·5–6·5 × 1·2–2·5 mm., minutely papillose; corolla lobes ovate, 7–15 × 4·5–10 mm., gradually narrowing into the pendulous tails; lobes (including the tails) 94–180 mm. long, pubescent outside except for the apex, glabrous inside. Stamens from 2·8 mm. exserted to 1 mm. included; filaments straight, with an abaxial swelling, 3·5–6·2 mm. long; anthers 5–7 × 1·3–2 mm., glabrous; acumen 0·1–0·4 mm. long. Ovary 1·2–2·3 × 1–2·5 mm., densely hispid with erect hairs up to 3·5 mm. long; style 11–18 mm. long; clavuncula 1·6–3 × 1·4–2·7 mm., stigma minute. Follicles divergent at an angle of 180°, tapering into an obtuse apex or ending in a knob, (15)20–38 cm. long and 1·5–3·2 cm. in diam.; exocarp pale brown, fairly thick and hard, shaggy with 4–18 mm. long villous protuberances, 0·5–1 mm. in diam. Seeds with the grain 11–24 × 2·5–5 mm., densely pubescent; rostrum glabrous for 18–60 mm. and bearing a coma for 30–50 mm.; coma 60–110 m. long.

Zambia. N: Kalambo Falls, 20.vi.1960, *Leach & Brunton* 10086 (K; MO; SRGH).

Also in Zaire and Tanzania. In leguminous woodland or *Acacia-Commiphora* bush, especially in rocky places. Flowering at the end of the dry and the beginning of the rainy season; mature fruits in the dry season.

3. **Strophanthus gardeniiflorus** Gilg in Engl., Monogr. Afr. Pfl.-Fam. & Gatt. **7**: 20 (1903). — Stapf in F.T.A. **4**, 1, Add.: 605 (1904). — White, F.F.N.R.: 351 (1962). — Beentje in Meded. Landb. Wag. **82**–4: 72, fig. 17, map 15 (1982). Type from Zaire.

Liana, 5–15 m. high, presumably evergreen; latex clear or milky. Trunk up to 5 cm. in diam., branchlets glabrous. Leaves petiolate; petiole 3–10(13) mm. long; lamina shiny and dark green above, dull and pale yellowish green beneath, elliptic or slightly obovate, 1·5–3·5(4) times as long as wide, 2–16·5 × 1–7·5 cm., cuneate at the base, rounded or acuminate at the apex (acumen 1–3 mm. long), slightly revolute at the margin, coriaceous, glabrous, with 5–9 pairs of secondary veins, with tertiary venation inconspicuous. Inflorescence 1–3-flowered, sessile or pedunculate, glabrous in all parts; pedicels 5–10 mm. long; bracts triangular or ovate, 1–9 × 1·5–4 mm., not sepal-like. Flowers fragrant. Sepals unequal, the inner larger than the outer, ovate or nearly circular, 6–12·5 × 3·5–7 mm., glabrous. Corolla tube 2·5–4·6 times as long as the calyx, 29–40 mm. long, widening gradually or at ⅖–½ of its length into a cylindrical or cyathiform upper part, at the mouth 12·5–22 mm. wide, glabrous outside, puberulous to pubescent inside; corona lobes narrowly triangular and sometimes undulate, 18–28 × 2·5–3 mm., sparsely pubescent; corolla lobes ovate, 1·7–2·4 times as long as wide, 24–46 × 11–21 mm., acute or acuminate, glabrous on both sides. Stamens 8–12 mm. exserted; filaments straight, 2·6–3·1 mm. long, anthers 26·5–30 × 1·3–2·2 mm., glabrous; acumen 17·5–20 mm. long. Ovary 1·9–2·8 × 1·9–3·1 mm. glabrous; style 16–19·5 mm. long; clavuncula 2·6–3·5 × 2·4–2·7 mm., stigma 0·3–0·7 mm. long. Follicles divergent at an angle of 180°, tapering into a narrow apex, 19–25 cm. long and 2·4–2·8 cm. in diam.; exocarp dark or purplish-brown, thick and hard, glabrous, lenticellate. Seeds with the grain 11·5–18 × 2–3·5 mm., densely puberulous or densely short-pubescent; rostrum glabrous for 4–11 mm. and bearing a coma for 10–28 mm.; coma erect or rarely reflexed, 48–60 mm. long.

Zambia. N: 20 km. on the Kawambwa-Mansa road, 30.x.1952, *Angus* 671 (BR; FHO; K; MO). W: Mwinilunga Distr., 6 km. N. of Kalene Hill Mission, 20.ix.1952, *White* 3301 (BR; FHO; K; WAG).
Also in Zaire and Angola. In riverine forest. Flowering in the second half of the dry season; mature fruits in the dry season.

4. **Strophanthus gerrardii** Stapf in Kew Bull. **1907**: 52 (1907). — Codd in Fl. Southern Afr. **26**: 292 (1963). — Beentje in Meded. Landb. Wag. **82**–4: 75, fig. 18, map 16 (1982). Type from S. Africa.

Liana, 3–12 m. high, deciduous, flowers appearing before or with the leaves; latex white. Trunk up to 2·5 cm. in diam., older branches with longitudinal corky ridges up to 6 cm. long and 1·5 cm. high, younger branches with triangular flat protuberances at the nodes; branchlets glabrous. Leaves petiolate; petiole 1–6 mm. long, lamina ovate, narrowly ovate, or elliptic, 1·7–4(5) times as long as wide, 3–6 × 1–2 cm., cuneate at the base, rounded or acuminate at the apex, rarely acuminate (acumen up to 3 mm. long), sometimes with a revolute margin, chartaceous or thinly coriaceous, glabrous with 3–6(8) pairs of secondary veins; tertiary venation inconspicuous. Inflorescence 1–2(5)-flowered, glabrous in all parts; pedicels 3–10 mm. long; bracts sepal-like. Sepals subequal, recurved for half their length, ovate or narrowly ovate, 2–6 times as long as wide, 3–10 × 1·5–2·5 mm., acute, glabrous. Corolla tube 14–25 mm. long, widening at ⅓–¼ of its length into a cylindrical or slightly infundibuliform upper part, at the mouth 6–13 mm. wide, glabrous outside and puberulous inside; corona lobes subulate, (1·3)2–5 × 1–1·6 mm. puberulous; corolla lobes ovate, 5–10 × 3–6 mm., gradually narrowing into the spreading tails; lobes (including the tails) 30–59 mm. long, glabrous on both sides but puberulous near the base. Stamens included for (0·6)2·7–6 mm., filaments straight or slightly curved, anthers 6·4–8·6 × 1–1·5 mm., glabrous; acumen 1·8–3·8 mm. long. Ovary 0·9–2 × 1·7–2·8 mm., glabrous; style 4·5–7 mm. long, smooth or wrinkled; clavuncula 1·4–2 × 1·3–1·9 mm., stigma 0·2–0·7 mm. high. Follicles divergent at an angle of 180–230°, tapering into a narrow obtuse apex, sometimes apically curved inwards at the apex, 11–24 cm. long and 1·5–2 cm. in diam.; exocarp brown or red-brown, thick and hard, glabrous, very densely lenticellate. Seeds with the grain 10–18 × 2·2–4 mm., densely pubescent; rostrum glabrous for 18–40 mm. and bearing a coma for 25–65 mm.; coma 45–85 mm. long.

Mozambique. M: 16 km. S. of Boane, 22.vii.1961, *Leach* 11197 (BM; G; K; LISC; M; MO; P; PRE; S; SRGH).
Also in S. Africa and Swaziland. In coastal "sand" forest or riverine forest, often in rocky places. Flowering from the beginning of the dry season until the rainy season; mature fruits throughout the year.

5. **Strophanthus holosericeus** K. Schum. & Gilg in Engl., Bot. Jahrb. **32**: 157 (1902). — Stapf in F.T.A. **4**, 1: 171 (1902). — Beentje in Meded. Landb. Wag. **82**–4: 91, fig. 22, map 20 (1982). Type from Zaire.

Liana, up to 9 m. long, deciduous, flowers appearing before or with the leaves. Branchlets densely puberulous or short-pubescent. Leaves petiolate; petiole 1–5 mm. long, lamina dark green above, silvery or yellowish beneath, ovate, 1–1·4 (in young leaves up to 2·5) times as long as wide, in mature leaves 6–17 × 5–12 cm., rounded or subcordate at the base, acuminate at the apex (acumen 1–5 mm. long), papyraceous or chartaceous, puberulous above, especially on the midrib and secondary veins, tomentellous beneath with 8–11 pairs of secondary veins, tertiary venation in older leaves, conspicuous beneath. Inflorescence on long or short branches or axillary, (1)2–9-flowered, sessile or pedunculate, lax or congested, densely pubescent in all parts; pedicels 3–6 mm. long; bracts sepal-like. Sepals subequal, ovate or narrowly elliptic, 10–17 × 3–6·5 mm., acute. Corolla tube 18–24 mm. long and widening at ⅗–⅘ of its length into a cyathiform upper part, at the mouth 11–14 mm. wide, densely short-pubescent outside and glabrous inside; corona lobes lingulate or subulate, 1·5–3 × 1–1·5 mm. glabrous; corolla lobes ovate, 6–10 × 6–9 mm., gradually or abruptly narrowing into the pendulous tails; lobes (including the tails) 100–150 mm. long, puberulous or pubescent outside. Stamens 0–3 mm. exserted; filaments 4–5 mm. long, straight; anthers 5·2–6·9 × 1·5–2 mm. glabrous; acumen 0·2–0·6 mm. long. Ovary 1·4–1·8 × 2 mm., densely hispid with erect hairs; style 13–16 mm. long; clavuncula 2 × 2 mm.; stigma minute. Follicles divergent at an angle of 170–180°, tapering towards a narrow apex and ending in a knob, 19–40 cm. long and 1·5–2·5 cm. in diam.; exocarp thick and hard, with a few or rather many small protuberances, puberulous or short-pubescent, fairly densely lenticellate, lenticels fairly elongate. Seeds with the grain 14–17 × 3–4 mm., densely pubescent; rostrum glabrous for 37–64 mm. and bearing a coma for 18–37 mm.; coma erect spreading, 45–75 mm. long.

Zambia. N: Kalambo Falls, 21.x.1947, *Greenway & Brenan* 8255 (FHO; K; NY).
Also in Zaire. In forest near waterfalls.

6. **Strophanthus hypoleucos** Stapf in Kew Bull. **1924**: 81 (1914). — White, F.F.N.R.: 351 (1962). — Beentje in Meded. Landb. Wag. **82**–4: 93, fig. 23, map 21 (1982). Type: Mozambique, Mt. M'Kota, *Stocks* 148 (K, holotype and isotype).

Shrub, 1–4 m. high, rarely a liana; deciduous, the flowers appearing together with the leaves; latex white. Branchlets densely tomentellous. Leaves petiolate; petiole 2–7·5 mm. long; lamina circular, elliptic, ovate, or obovate, 0·9–2 times as long as wide (in young leaves up to 3 times as long as wide), 1·7–8 × 1·3–5(8·5) cm., cuneate, rounded or subcordate at the base, slightly emarginate, rounded, or acute at the apex (rarely with an acumen up to 3 mm. long), papyraceous or thinly coriaceous, densely short-pubescent or tomentellous above, especially on the midrib and veins, and tomentose beneath with whitish hairs, with 4–8 pairs of secondary veins; tertiary venation inconspicuous. Inflorescence 1–6-flowered, pedunculate, congested and sometimes partly reduced, densely tomentellous in all parts; pedicels 4–15 mm., long; bracts sepal-like. Sepals subequal, reddish-brown, elliptic, broadly elliptic, or obovate, 5·5–12·5 × 2–9 mm., acute, densely tomentellous. Corolla tube 14–22 mm. long and widening at slightly less than ½–⅔ of its length into a more or less shallowly cyathiform upper part, at the mouth 7–14 mm. wide, densely short-tomentellous or pubescent outside, glabrous inside; corona lobes triangular, 0·6–2 × 1–1·8 mm., minutely papillose; corolla lobes ovate, 6–12 × 5·5–11 mm., fairly abruptly narrowing into the spreading or pendulous tail; lobes (including the tail) 22–57 mm. long, densely short-pubescent or tomentellous outside, glabrous inside. Stamens from 3 mm. exserted to 0·4 mm. included; filaments straight, 2·3–4·5 mm. long; anthers 4·3–5·3 × 0·7–1·8 mm., glabrous; acumen 0–0·2 mm. long. Ovary 1–1·3 × 1·5–2·5 mm. densely pubescent; style 9–14·5 mm. long; clavuncula 1·3–2·2 × 1·3–2·2 mm.; stigma minute. Follicles divergent at an angle of 180–200°, tapering towards the apex and ending in a small knob, 12·5–23 cm. long and 2 cm. in diameter; exocarp chocolate-brown, thick and hard, short-pubescent or glabrescent, sparsely or densely lenticellate. Seeds with the grain 8–12·5 × 2·8–3·2 mm., densely pubescent; rostrum glabrous for 15–30 mm. and bearing a coma for 12–33 mm.; coma 32–53 mm. long.

Mozambique. N: Montepuez, 17.x.1942, *Mendonça* 906 (BM; LISC). M: Imala, Mecuburi road, 27.xi.1936, *Torre* 1049 (COI; LISC; PRE). Z: Mts. do Ile, Errego, 22.vi.1943, *Torre* 5569 (BM; LISC).
Also in Tanzania. On rocks, in woodland.
Flowering towards the end of the dry and the beginning of the rainy season; mature fruits in the dry season.

7. **Strophanthus kombe** Oliver in Hooker, Ic. Pl., Ser. 3, **1**: 79, pl. 1098 (1871). Stapf in F.T.A. **4**, 1: 173 (1902). — Codd in Bot. Surv. Mem. **26**: 158 (1951). — White, F.F.N.R.: 351 (1962). — Codd in Fl. Southern Afr. **26**: 290 (1963). — Verdcourt & Trump, Comm. Pois., Pl. E. Afr.: 148, fig. 10 (1969). — Beentje in Meded. Landb. Wag. **82**–4: 96, fig. 24–25, map 22 (1982). TAB. **112**. Type: Malawi, Manganja Hills, ix.1861, *Meller* (K, holotype).
    S. hispidus var. *kombe* (Oliver) Holmes in Pharm. Journ. Trans., Ser. 3, **21**: 223 (1890). Type as above.

Sarmentose shrub, 1–3·5 m. high, or liana, 2·5–20 m. high, deciduous, flowers and leaves appearing at the same time; latex clear, white, or yellow. Roots thick and

Tab. 112. STROPHANTHUS KOMBE. 1, flowering branches with young leaves (× ⅔), from *Strid* 2315; 2, mature leaf (× ⅔), 3, detail of lower leaf surface (× ⅔), 2–3 from *Pienaar* 212; 4, flower (× 2), from *Chase* 8056; 5, section of flower, (× 2), 6, stamens and apex of gynoecium (× 8), 7, ovary (× 8), 5–7 from *Davies* 2222; 8, follicle (× ⅓), 9, detail of exocarp (× ⅔), 10, seed with detail of grain (× ⅔), 8–10 from *Chase* 8056.

fleshy. Trunk up to 10 cm. in diam.; bark reddish-brown or grey-brown; branches scabrous as the result of persistent hair-bases; branchlets densely hispid. Leaves petiolate; petiole 1·5–5 mm. long; lamina dark green, paler beneath, ovate or elliptic, less often obovate or nearly circular, 1·1–2·3 times as long as wide, in mature leaves 8–23·5 × 5–16·5 cm. cuneate, rounded, or subcordate at the base, obtuse, acute, or acuminate at the apex (acumen 1–11 mm. long), papyraceous or chartaceous, in young leaves densely hispid on both sides, in older leaves glabrescent above, with 7–13 pairs of secondary veins; tertiary venation conspicuous beneath. Inflorescence 1–12-flowered, pedunculate, fairly congested, densely hispid in all parts; pedicels 3–14(20) mm. long; bracts sepal-like. Flowers fragrant. Sepals subequal, narrowly ovate or linear, 9–20(27) × 1·5–3·5 mm., acute, densely hispid. Corolla tube 13–24 mm. long and widening at almost ½–⅔ of its length into a cyathiform upper part, at the mouth (6)8–14 mm. wide, densely hispidulous outside except for the base, sparsely hispidulous inside except for the base; corona lobes lingulate, 1–3 × 1–2·4 mm., minutely puberulous or papillose; corolla lobes ovate, 3–16 × 4–8·5 mm., gradually or fairly abruptly narrowing into the pendulous tails; lobes (including the tails) 100–160(–200) mm. long, puberulous except for the inner side of the tails. Stamens included for 2·7–7·3 mm.; filaments curved, 0·6–1·2(2) mm. high; anthers 3·7–6·2 × 0·6–1·1 mm., glabrous; acumen 0·1–0·5 mm. long. Ovary 0·8–1·7 × 1·5–2·3 mm., densely hispid with long erect hairs, sometimes glabrous at the base; style 6·5–13·5 mm. long; clavuncula 1–1·8 × 0·9–1·3 mm.; stigma minute. Follicles divergent at an angle of 180°, long tapering toward the apex and ending in a small or large knob, rarely without knob and then with an obtuse apex, 15–47 cm. long and 1·3–2·6 cm. in diam.; exocarp thick and hard, densely hispid or pubescent in young fruits and glabrescent when maturing, especially on the adaxial side densely lenticellate. Seeds with the grain 11–21 × 2·5–4·5 × 1·5 mm., densely pubescent; rostrum glabrous for 20–57 mm. and bearing a coma for 20–42 mm.; coma 42–80 mm. long.

Botswana. N: Kazungula, confluence of Linyanti R. and Zambezi R., vi.1886, *Holub* 3225 (W; Z). **Zambia.** C: Luangwa Game Reserve, near Mfuwe, 19.x.1967, *Astle* 5121 (K; SRGH). S: Gwembe Distr., Chete Gorge, 17.xi.1955, *Bainbridge* 193/55 (FHO; SRGH). S: Victoria Falls Knife Edge near Livingstone, i.1914, *Rogers* 13060 (BOL). **Zimbabwe.** N: Sebungwe Distr., Kariyangwa, 13.xi.1956, *Lovemore* 492 (BR; K; PRE; SRGH). W: 3 km. W. of Lukozi R. bridge on Bulawayo-Victoria Falls Road, 3.xi.1973, *Raymond* 191 (B; BR; E; K; PRE; SRGH). S: Ndanga Distr., Chipinda Pools, 18.x.1951, *McGregor* 82/51 (FHO; MO; NY; PRE; SRGH). **Malawi.** S: Lengwe Nat. Park, 6.x.1970, *Hall-Martin* 1398 (K; SRGH). **Mozambique.** N: Pemba (Porto Amelia), 4.x.1964, *Gomes e Sousa* 4843 (SRGH). Z: between Marral and Quelimane, 15.x.1941, *Torre* 3675 (BM; LISC). T: Cahora Bassa, road between dam and Songo, 27.x.1973, *Correia et al.* 3645 (WAG). MS: Sofala Prov., Jambara, N. of Chemba, 28.x.1972, *Bond* 21 (LISC; SRGH). GI: Inhambane Prov., 10 km. N. of Mavume, x.1938, *Gomes e Sousa* 2168 (K; LISC; PRE). M: Porto Henrique, 27.vii.1907, *Gerstner* 6630 (BOL; PRE).

Also in Kenya, Tanzania and S. Africa. Locally frequent at low altitudes in woodland in hot, low-rainfall areas, often on inselbergs.

Flowering towards the end of the dry and the beginning of the rainy season; mature fruits in the dry season.

8. **Strophanthus luteolus** Codd in Bothalia **7**: 454 (1961). — Codd in Fl. Southern Afr. **26**: 289 (1963); in Flow. Pl. Afr.: t. 1561 (1969). — Beentje in Meded. Landb. Wag. 82–4: 105, fig. 27, map 24 (1982). Type from S. Africa.

Liana, 2–6 m. high; deciduous, flowers and leaves appearing at the same time; latex white or yellow. Roots semi-tuberous; trunk up to 1·25 cm. in diam.; branchlets puberulous. Leaves petiolate; petiole 1–5 mm. long; lamina elliptic, obovate, or rarely ovate, 1·4–3(4·6) times as long as wide, 2·5–6·5 × 1·1–3·1 cm., cuneate at the base, rounded, mucronate, acute, or acuminate at the apex (acumen 1–3(7) mm. long), membranaceous or papyraceous, glabrous or sparsely to densely puberulous, with 4–6(8) pairs of secondary veins; tertiary venation conspicuous beneath. Inflorescence 1(6)-flowered, sessile or pedunculate, fairly congested if branched, puberulous in all parts; pedicels 3–13 mm. long; bracts sepal-like. Sepals subequal, elliptic or narrowly ovate, 5–14·8 × 1–4 mm., acute, puberulous. Corolla tube 10–18 mm. long and widening at ½–⅔ of its length into a cyathiform upper part, at the mouth 6–12 mm. wide, puberulous on both sides except for the base inside; corona lobes lingulate or subulate, 1·5–3·5 × 1–2 mm. minutely papillose or minutely puberulous;

corolla lobes ovate, 4–8 × 2·8–5·5 mm., abruptly narrowing into pendulous tails; lobes (including the tails) 36–86 mm. long, puberulous on both sides except for the apex. Stamens included for 0·5–2 mm.; filaments curved, 0·9–1·1 mm. high, puberulous or pubescent; anthers 3·3–4·5 × 1–1·3 mm., glabrous, except for a small pubescent patch around the filament; acumen 0·2–0·3 m. long. Ovary 0·7–1·8 × 1·1–2·4 mm., pubescent in the upper part or all over with erect hairs; style 7–9·6 mm. long; clavuncula 0·9–1·1 × 0·8 × 1·2 mm.; stigma minute. Follicles divergent at an angle of 180–200°, long-tapering towards the apex and ending in a small knob, 16–29 cm. long and 1–1·5 cm. in diam.; exocarp brown, fairly thick and hard, densely pubescent in young fruits and glabrescent in older fruits, densely lenticellate. Seed with the grain 12–17 × 2–3 mm., densely short-pubescent; rostrum glabrous for 28–45 mm. and bearing a coma for 30–45 mm.; coma 58–80 mm. long.

Mozambique. M: 3 km. S. of Bela Vista, turnoff on Catuane road, 15.x.1963, *Leach & Bayliss* 11949 (K; SRGH; Z).
Also in S. Africa. In "sand" forest on dunes near the coast.
Flowering towards the end of the dry and the beginning of the rainy season; mature fruits probably in the dry season.

9. **Strophanthus nicholsonii** Holmes in Pharm. Journ., Ser. 4, **5**: 209 (1897). — Stapf in F.T.A. **4**, 1: 172 (1902). — White, F.F.N.R.: 352 (1962). — Beentje in Meded. Landb. Wag. 82–4: 113, fig. 30, map 27 (1982). Type: Zambia, between Lusangadzi and upper Luangwa Rs., *Nicholson* s.n. (PHA, holotype; P, PRE, isotypes).

Shrub, 0·5–2·5 m. high, densely branched, sometimes scandent and up to 6 m. high, deciduous; flowers appearing before or with the leaves; latex clear or orange. Branchlets densely puberulous. Leaves petiolate; petiole 1–2(3) mm. long; lamina medium to dark green above, pale yellowish-green beneath, obovate, rarely almost circular, 1·2–2·5 times as long as wide, 2·5–5·7 × 1·3–3·7 cm., rounded or cuneate at the base, rounded or acute at the apex, papyraceous, densely puberulous above, shortly tomentose beneath, with 4–9 pairs of secondary veins; tertiary venation inconspicuous. Inflorescence 1–5(10)-flowered, sessile or pedunculate, densely puberulous in all parts; pedicels 2–7 mm. long; bracts sepal-like. Flowers fragrant. Sepals subequal, linear or narrowly ovate, 4·5–12 × 1·3–3 mm., acute, densely puberulous. Corolla tube 10–19 mm. long and widening at slightly more than ½–¾ of its length into an infundibuliform upper part, at the mouth 5–14 mm. wide, puberulous outside and puberulous or glabrous inside; corona lobes lingulate, 1–2·5 × 1–1·6 mm., glabrous or minutely papillose; corolla lobes ovate, 4–15 × 3·8–8 mm., narrowing into the pendulous tails; lobes (including the tails) 55–115 mm. long, puberulous outside and glabrous or puberulous inside. Stamens from 1 mm. exserted to 2·1 mm. included; filaments straight, 1–2·3 mm. long; anthers 4–4·8 × 1–1·2 mm., glabrous; acumen 0·1–0·2 mm. long. Ovary 0·8–1·5 × 1·2–1·8 mm. densely pubescent; style 6–10 mm. long; clavuncula 1–1·8 × 0·9–1·3 mm.; stigma minute. Follicles of unknown divergence, long-tapering towards the narrow apex and ending in a small knob or an obtuse apex, 17–28 cm. long and 1·4–2·7 cm. in diam.; exocarp dark brown or purplish-brown, thick and hard, pubescent in young fruits and later glabrescent, fairly densely lenticellate; lenticels elongate. Seeds with the grain 12–22 × 3–5 × 1–1·5 mm., densely pubescent or lanate; rostrum glabrous for 13–44 mm. and bearing a coma for 20–65 m., coma 35–80 mm. long.

Zambia. E: 27 km. N. of Jumbe, 11.x.1958, *Robson* 53 (BM; BR; K; LISC; PRE; SRGH). C: Katondwe Mission, 13.xi.1966, *Fanshawe* 9839 (K). S: Gwembe, 1.xi.1955, *Bainbridge* 171/55 (FHO; K; SRGH). **Zimbabwe.** N: Urungwe Distr., Gache-Gache Triangle, 21.xi.1953, *Wild* 4266a (B; BR; FI; K; MO; PRE; S; SRGH). W: 15 km. from Mbala Lodge on Hwange (Wankie) road, 22.x.1968, *Rushworth* 1222 (K; LISC; PRE; SRGH). E: Inyanga N. Res., Gairezi R. bank, x.1958, *Davies* 2518 (K; PRE; SRGH). **Malawi.** N: Mzimba Distr., above Lake Kazuni, 17.x.1973, *Pawek* 7406 (MO; PRE; UC; WAG). **Mozambique.** T: Zambezi R., 60 km. W. of Msasa, 10.i.1951, *Chase* s.n. (MO; NY).
Not known from elsewhere. In mopane woodland, alt. 400–1100 m. Flowering at the end of the dry season; mature fruits in the dry season.

10. **Strophanthus petersianus** Klotzsch in Peters, Reise Mossamb., Bot. **1**: 276 (1861). — Stapf in F.T.A. **4**, 1: 182 (1902). — Codd in Bot. Surv. Mem. **26**: 158, fig. 147 (1951). — Codd in Fl. Southern Afr. **26**: 291 (1963). — Verdcourt & Trump, Comm. Pois. Pl. E.

Afr.: 136 (1969). — Retief in Fl. Pl. Afr. **42**: t. 1658 (1973). — Beentje in Meded. Landb. Wag. **82**–4: 120, fig. 33, map 30 (1982). Type: Mozambique, Tete, Zambezi R., *Peters* s.n.
    S. petersianus var. *grandiflorus*. N.E. Br. in Kew Bull. **1892**: 126 (1892). Type: Mozambique. Maputo, Delagoa Bay, *Monteiro* 1 (K, holotype; FI; G; P; W, isotypes).
    S. grandiflorus (N.E. Br.) Gilg in Engl. Monogr. Afr. Pfl.-Fam. & Gatt. **7**: 161, t. 7 (1903). Type as above.
    S. sarmentosus var. *verrucosus* Pax in Engl. Bot. Jahrb. **15**: 374 (1892). Type from Kenya.

Sarmentose shrub or liana, 1–15 m. high, deciduous, flowers appearing with or rarely before the leaves; latex — if present — white or reddish. Trunk up to 10 cm. in diam., pale grey; branches at the nodes (or less often in between), with 2–4 corky laterally compressed triangular protuberances up to 25 mm. high; branchlets glabrous or rarely puberulous. Leaves petiolate; petiole (2)3–13 mm. long; lamina dark green, paler beneath, elliptic or ovate, 1·3–3(3·4) times as long as wide, 2·8–11 × 1·7–5·2 cm., cuneate or rounded at the base or decurrent into the petiole, acuminate at the apex (acumen 2–10 mm. long, obtuse), sometimes recurved at the margin, papyraceous or less often thinly coriaceous, glabrous or exceptionally sparsely puberulous, with 4–6(8) pairs of secondary veins; tertiary venation conspicuous, especially beneath. Inflorescence 1–2(4)-flowered, sessile or pedunculate, glabrous or occasionally puberulous in all parts; pedicels 3·5–11 mm. long; bracts sepal-like or subscarious. Flowers fragrant. Sepals erect or spreading, fairly unequal, ovate or narrowly elliptic, 5–21 × 2–5·5 mm., acute, glabrous or exceptionally puberulous. Corolla tube (13)15–37 mm. long and widening at ⅓–⅔ of its length into a cyathiform upper part, at the mouth 10–29 mm. wide, glabrous outside and puberulous inside; corona lobes narrowly triangular and often undulate, 6–15 × 1·7–4 mm., glabrous; corolla lobes ovate, 9–16 × 6–15 mm., gradually narrowing into the pendulous tails; lobes (including the tails) 90–205 mm. long, glabrous on both sides. Stamens included for 1·5–11 mm., rarely 0–1·5 mm. exserted; filaments straight or nearly so, 2·6–5·2 mm. long; anthers 6–10 × 1·1–2 mm., glabrous; acumen 1–4 mm. long. Ovary 0·8–2·6 × 1·5–2·8 mm., glabrous; style 7·5–14·5 mm. long; clavuncula 1·8–2·8 × 1·3–2·3 mm.; stigma minute. Follicles divergent at an angle of 180°, tapering towards the apex and ending in a narrow obtuse point or in a small knob, 20–37 cm. long and 2·2–3·5 cm. in diam.; exocarp dark brown, thick and hard, glabrous, sparsely or densely lenticellate, rarely not lenticellate. Seeds with the grain 10–18 × 2·8–4 × 1 mm., densely pubescent; rostrum glabrous for (20–)35–65 mm. and bearing a coma for 10–52 mm.; coma (38)60–90 mm. long.

Zambia. S: Mazabuka, 2.ii.1958, *Drummond* 5447 (SRGH). **Zimbabwe.** N: Kariba Gorge slopes, x.1959, *Goldsmith* 46/59 (BR; K; L; MO; PRE; SRGH). W: Hwange (Wankie), iv.1954, *Levy* 1008 (E; PRE; SRGH). **Malawi.** S: Shire Highlands, Chiromo, *Scott-Elliot* 2793 (BM; K). **Mozambique.** N: Angoche, Metangula beach, 16.x.1965, *Mogg* 32421 (LISC; SRGH). T: Cahora Bassa, Posto de Milicias Rio, 15.xi.1973, *Correia et al.* 3833 (WAG). MS: Manica Prov. lower slopes of Mt. Zembe, 21.vi.1959, *Leach* 9126 (K; PRE; SRGH). GI: Inhambane, x.1936, *Gomes e Sousa* 1898 (BR; COI; FI; K; LISC). M: Moamba area, 171.x.1980, *P. Jansen et al.* 7553 (WAG).
    Also in Kenya, Tanzania and S. Africa. In coastal forest and woodland, often on rocky places; alt. 0–1100 m.
    Flowering towards the end of the dry and beginning of the rainy season; mature fruits in the dry season.

11. **Strophanthus speciosus** (Ward & Harvey) Reber in Fortschr. (Genf) **3**: 299 (1887). — Stapf in F.C. **4**, 1: 511 (1907). — Marloth in Fl. S. Afr. **3**: pl. 18 (1932). — Codd in Fl. Southern Afr. **26**: 293, fig. 42 (1963). — Coates-Palgrave, Trees of Southern Afr.: 797 (1977). — Beentje in Meded. Landb. Wag. **82**–4: 143, fig. 39, phot. 6, map 36 (1982). Type from S. Africa.
    Christya speciosa Ward & Harvey in Journ. Bot., Lond. **4**: 134 (1842), excl. specim. *Ecklon & Zeyher*. Type as above.
    S. capensis A. DC., Prodr. **8**: 419 (1844). Type from S. Africa.

Shrub, 1–4 m. high, or liana, up to 16 m. high, presumably evergreen; latex clear or white. Trunk up to 3 cm. in diam., branching trichotomously; branchlets glabrous. Leaves petiolate, ternate, rarely opposite or quaternate; petiole 2–12 mm. long; lamina glossy and medium or dark green above, dull and paler beneath, narrowly elliptic or slightly obovate, 2·5–6(9) times as long as wide, 2·2–11·5 × 0·6–3·4 cm., decurrent into the petiole, acute, acuminate, or rarely rounded at the apex (acumen 1–6 mm. long), often with a slightly revolute margin, coriaceous or thinly so, glabrous, with 6–19 pairs of secondary veins; tertiary venation sometimes conspicuous. Inflorescence 3–26-flowered, pedunculate, congested, glabrous in all parts or puberulous in some; pedicels 6–21 mm. long; bracts sepal-like. Sepals subequal, narrowly ovate, 3–14·5 × 1·2–2(2·7) mm., acute, with some hairs near the apex or puberulous all over. Corolla tube (6·5)9–14 mm., long and widening at $\frac{1}{3}$–$\frac{1}{2}$ of its length into a cyathiform upper part, at the mouth 5·5–11 mm. wide, minutely puberulous near the apex outside, minutely papillose inside; corona lobes subulate, 1·8–4·9 × 1–2·2 mm., minutely papillose; corolla lobes ovate, 2–6 × 3–5 mm., gradually narrowing into the spreading tails; lobes including the tails 19–50 mm. long, puberulous outside and sometimes also near the base inside. Stamens included for 0–3·5 mm.; filaments straight, 0·5–0·8 mm., long; anthers 3·4–6 × 0·4–1 mm., pubescent in the upper third or for more than half of their length; tails 0·2–0·7 mm. long; acumen 0·9–2·5 mm. long. Ovary 0·8–1·8 × 1·2–2·1 mm., puberulous; style 3·9–5·5 mm. long; clavuncula 0·8–1·6 × 0·7–1·9 mm., pubescent at the apex; stigma minute. Fruit 1, or occasionally 2, in a single infrutescence; follicles divergent at an angle of (60)100–230°, tapering towards the narrow apex and ending in a minute knob, with the extreme apex sometimes curved inwards, (7)10–22 cm. long and 0·7–1·5 cm. in diam.; exocarp fairly thin, hard, glabrous, not lenticellate. Seeds with the grain 13–22 × 2–4·5 mm. densely puberulous; rostrum glabrous for 0–2 mm. and bearing a coma for 7–15 mm.; coma 22–43 mm. long.

Zimbabwe. C: Wezda Mt., 23.ii.1964, *Wild* 6340 (K; LISC; PRE; SRGH). E: Vumba Mts. near Mutare (Umtali), *Obermeyer* 2038 (PRE).

Also in Swaziland and S. Africa. In evergreen forest, and ravine or scrub forest; alt. (300)900–1500(1800) m.

Flowering towards the end of the dry season; mature fruits throughout the year, with a peak in the dry season.

12. **Strophanthus welwitschii** (Baill.) K. Schum. in Engl. & Prantl, Pflanzenfam. **4**, 2: 59 (1900). — White, F.F.N.R.: 352 (1962). — Verdcourt & Trump, Comm. Pois. Pl. E. Afr.: 135 (1969). — Beentje in Meded. Landb. Wag. **82**–4: 154, fig. 43, phot. 7, map 40 (1982). Type from Angola.
    Zygonerion welwitschii Baill. in Bull. Mens. Soc. Linn. Paris **1**: 758 (1888). Type as above.
    S. ecaudatus Rolfe in Bol. Soc. Brot. **11**: 85 (1893). Type from Angola.
    S. parvifolius K. Schum. in Engl. & Prantl., Nat. Pflanzenfam. **4**, 2: 182 (1895). Type: the description.
    S. verdickii De Wild. in Ann Mus. Congo Belge, Bot., Sér. 4, **1**: 103, t. 31 f. 1–6 (1903). Type from Zaire.
    S. verdickii var. *latisepalus* De Wild., tom. cit.: 104, t. 3 f. 7–13 (1903). Type from Zaire.
    S. gilletii De Wild., tom. cit.: 105 (1903). Type from Zaire.
    S. katangensis Staner in Ann. Soc. Sci. Bruxelles, Sér. B, **52**: 94 (1932). Type from Zaire.

Sarmentose shrub or small tree, 0·60–5 m. high, or liana, up to 8 m. high, deciduous; latex clear or white. Trunk up to 10(40) cm. in diam., dark brown or grey; branchlets minutely puberulous. Leaves petiolate opposite or rarely in some branches ternate or quaternate; petiole 1–5 mm. long, lamina dull and medium or dark green above, pale yellowish- or whitish-green beneath, ovate, narrowly elliptic, or rarely slightly obovate, 1·2–4 (on long shoots up to 6) times as long as wide, mature leaves up to 8·5 × 4·2 cm., cuneate or nearly rounded at the base, rounded, acute, or acuminate at the apex (acumen up to 8 mm. long), slightly revolute at the margin and there often reddish beneath, thinly coriaceous, glabrous or sparsely puberulous, especially on the midrib and near the margins, with 3–8(10) pairs of secondary veins; tertiary venation sometimes conspicuous beneath. Inflorescence 1–2(5)-flowered, sessile or pedunculate, glabrous or puberulous; pedicels 3–9(17) mm. long; bracts linear or narrowly ovate, subscarious. Flowers fragrant. Sepals erect or with the upper half recurved, subequal, ovate or narrowly ovate, 5–19 × 2–5·5(7) mm., acute, puberulous near the base or less often glabrous or puberulous all over. Corolla tube (13·5)17–38 mm. long and widening gradually at $\frac{1}{5}$–$\frac{2}{5}$ of its length into a cyathiform upper part, at the mouth (11)15–27(–37) mm. wide, puberulous outside near the base, minutely pubescent or puberulous inside; corona lobes narrowly triangular, 5–23 × 1·5–5 mm., minutely papillose or puberulous; corolla lobes spreading or recurved, ovate, (10)14–38(48) × (7·5)10–24(29) mm., acute, minutely puberulous inside. Stamens included for 2–12 mm.; filaments straight or nearly so, 3–5·5 mm. long; anthers 5–8·2 × 1·2–1·9(2·5) mm., glabrous; acumen 0·2–1·7 mm. long. Ovary 0·9–2·5 × 1·4–2·9 mm., glabrous; style 8–14·5 mm., long; clavuncula (1·3)2–3·1 × 1·5–2·3 mm.; stigma minute. Follicles divergent at an angle of 160–240°, long-tapering towards a narrow apex and ending in an obtuse point, or rarely in a small knob, with the extreme apex sometimes curved inwards, 10·5–33·5 cm. long and 1–2·5 cm. in diam.; exocarp dark brown or purple-brown, thick and hard, glabrous, densely lenticellate. Seeds with grain 8·5–19·5 × 2·5–4 × 1 mm., densely pubescent; rostrum glabrous for 17–54 mm., and bearing a coma for 22–55 mm., coma 32–95 mm. long.

Zambia. B: Kalabo Boma, 13.ii.1952, *White* 2068 (BR; FHO; K; MO). N: Isoka Boma, 3.x.1936, *Burtt* 6286 (BM; BR; K). W: 133 km., Solwezi-Mwinilunga road, 16.ix.1952, *White* 3258 (BR; FHO; K; MO; WAG). C: Kafulafuta Junction, 12.x.1961, *Linley* 196 (BR; K; M; SRGH).

Also in Zaire, Tanzania and Angola. In miombo woodland, often on rocky places or in riverine forest; alt. 300–1800 m.

Flowering towards the end of the dry and the beginning of the rainy season; mature fruits in the dry season.

23. WRIGHTIA R. Br.

Wrightia R. Br. in Mem. Wern. Soc. **1**: 73 (1811). — Pichon in Not. Syst. **14**: 77 (1951). — Ngan in Ann. Miss. Bot. Gard. **52**: 133 (1965) non Solander & Naudin (1852).
Balfouria R. Br., tom. cit.: 70 (1811).
Anasser Blanco, Fl. Filipp., ed. 1: 112 (1837), non Jussieu (1789).
Piagiaea Chiov., Fl. Somala **II**: 290 (1932).
Wallida Pichon, tom. cit.: 87 (1951).
Scleranthera Pichon, tom. cit.: 88 (1951).

Shrubs or trees, occasionally climbers (not in F.Z. area). Leaves opposite, with colleters in the axils, petiolate; lamina eglandular. Inflorescence terminal, aggregated, dichasial or monochasial, few- to many-flowered. Flowers actinomorphic except for the subequal sepals. Sepals almost free, imbricate, with 1–2 alternate colleters. Corolla subrotate to subinfundibuliform, occasionally infundibuliform or hypocrateriform; tube cylindrical to campanulate, constricted at or near the mouth or not; lobes overlapping to the left; corona variously shaped, only absent in the Asian *W. religiosa*. Stamens exserted or included (not in F.Z. area); anthers narrowly triangular, partly fertile, introrse. Ovary composed of 2 basally free or slightly connate carpels, united at the apex by the style; style often split at the base and widened at the apex; clavuncula coherent with the anthers, subcapitate or subcylindrical, with a basal collar; stigma bi-apiculate, minute. Disk none. Fruit composed of two follicles; follicles connate at the extreme base and sometimes also at the apex (not in F.Z. area), terete to laterally compressed, dehiscent throughout by an adaxial slit. Seeds numerous, narrowly fusiform, not rostrate, with an apical coma directed towards the base of the fruit.

An Old World genus of 23 species, two of which occur in Africa.

Wrightia natalensis Stapf in Kew Bull. **1907**: 51 (1907). — Codd in Fl. Southern Afr. **26**: 296 (1963). — Ngan in Ann. Miss. Bot. Gard. **52**: 161 (1965). TAB. **113**. Type from S. Africa (Natal).

Tree 3–15 m. high, repeatedly dichotomously branched. Bark pale grey-brown, inner bark white or brownish, with white latex. Branches lenticellate; branchlets glabrous or only sparsely pubescent when very young, terete, often sulcate when dry. Leaves shortly petiolate; petioles glabrous or sparsely pubescent, 3–5 mm. long, with colleters in the axils; lamina membranous when dry, narrowly ovate or narrowly elliptic, 3–6·5 times as long as wide, 3·5–10 × 1–2·7 cm., long-acuminate, cuneate or rounded at the base, glabrous or sometimes with minute scattered hairs above, pubescent at the base of the midrib or sometimes entirely glabrous beneath; secondary veins 8–15 at each side, fairly inconspicuous. Inflorescence solitary, terminal, often in the forks or axillary, erect 2–5 × 2–5 cm. Peduncle 3–5 mm., long, slender, pubescent as the branches. Bracts linear, about as long as the sepals,

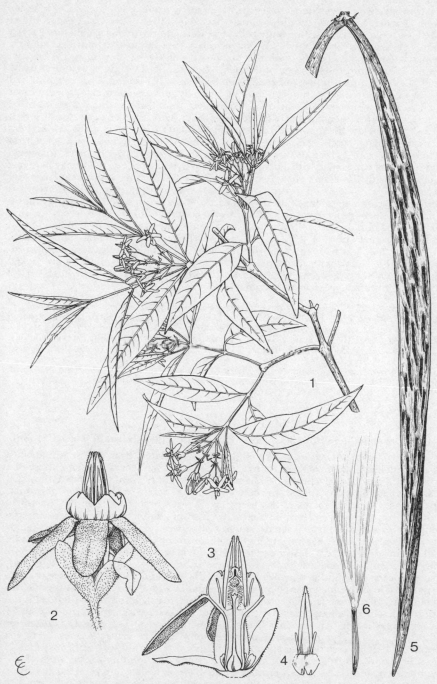

Tab. 113. WRIGHTIA NATALENSIS. 1, flowering branch ($\times \frac{2}{3}$); 2, flower ($\times 4$); 3, flower in longitudinal section ($\times 4$); 4, part of corona with anther ($\times 4$); 5, fruit ($\times \frac{2}{3}$); 6, seed ($\times 1$); 1–4 from *de Winter & Vahrmeijer* 8464; 5–6 from *Stephen* 712.

deciduous. Pedicel 3–5 mm. long, pubescent. Sepals pale green, with a membranous margin, persistent even beneath the fruit, narrowly ovate or elliptic, 2–2·4 times as long as wide, 5–6 × 2·2–2·5 mm., obtuse, sparsely pubescent outside, especially at the base, ciliate, inside with 5 large colleters alternating with them; colleters 1 × 0·8 mm., flat, entire or 3-lobed. Corolla creamy; tube almost cylindrical, slightly constricted at the insertion of the stamens, slightly shorter than the sepals, 4–5 × 1·7–2·8 mm., glabrous on both sides; lobes elliptic, 2–3·3 times as long as wide, 7–9 × 2·7–3·5 mm., obtuse, puberulous on both sides, spreading. Corona composed of 2 rings inserted at the mouth; outer of 5 basally united emarginate lobes, almost half as long as the anthers, each with triangular apices, and inside it a series of 10 shorter filiform appendages near the edges of the anthers. Stamens exserted, inserted 0·5–1 mm. below the mouth of the corolla; filaments very short, glabrous; anthers narrowly triangular, 4·5–5·5 × 1–1·4 mm., acuminate at the sterile apex, auriculate at the base, introrse, glabrous outside or at the apex with a few stiff hairs, inside on lower half of connective with a penicillate line of hairs just below the level where it is coherent with the clavuncula. Pistil glabrous, 6·8–7·7 mm., long; ovary broadly ovoid, 1·5–1·7 × 1·5–2 × 1·5–2 mm., composed of two free carpels; style often split at the base, 3·5–4·5 mm. long, cylindrical and at the apex with a globose head, c. 1 mm. diam. supporting the clavuncula; clavuncula composed of an entire recurved ring 0·4 × 1–1·2 mm. in size, followed by a small cone topped by a globe 0·3–0·6 mm. in diam.; stigma minute, 0·2 × 0·1 mm.; whole of clavuncula and stigma 1·4–1·5 × 1–1·2 mm. Ovules c. 100 in each cell. Follicles slender, cylindrical, 20–40 × 0·8–1 cm., with many pale lenticels, dehiscent throughout by an adaxial longitudinal slit and becoming flat and 15–17 mm. wide. Seed pale brown, 15–19 × 2 mm., longitudinally striate, with coma 3–5 cm. long.

Zimbabwe. E: Chipinge (Chipinga) Distr., Sabi Valley, Chisumbanje, Masagwe R., fl. 21.x.1965, *Plowes* 2724 (SRGH). S: Mwenezi Distr., Gona-re-Zhou National Park, between Sango (Vila Salazar) and Guluene R., 5 km. from Mozambique border, fl. 18.x.1975, *Drummond* 10403 (B; M; MO; SRGH). **Mozambique.** GI: "Piccadilly Circus" (?), veg. 21.ii.1969, *Sherry* 13/69 (LISC; SRGH). M: Santaca, fl. & dry carpels 14.viii.1948, *Gomes e Sousa* 3791 (COI; PRE; S).

Also known from Swaziland and S. Africa (Transvaal, Natal). Dry woodland and scrub forest on hillsides in light sandy soil, medium to low rainfall areas; 0–1000 m.

24. PLEIOCERAS Baill.

Pleioceras Baill. in Bull. Soc. Linn., Paris **1**: 759 (1888); Hist. Pl. **10**: 210 (1889).

Shrubs, lianas or trees up to 9 m. high, with white latex. Branches unarmed, with pale lenticels; branchlets terete, often sulcate when dry. Leaves opposite, those of a pair equal or subequal, shortly petiolate, with colleters in one or two rows in the axils; lamina acuminate, rounded or cuneate at the base. Inflorescences terminal, paniculate, several- to many-flowered, lax. Lower bracts leafy; others narrowly oblong, acute, with colleters in the axils. Flowers actinomorphic except for the sometimes slightly unequal sepals, small. Calyx persistent beneath the fruit; sepals green, connate at the base, ovate, imbricate, entire, pubescent or glabrous outside, ciliate, inside at the base near the edge with 0–1 triangular or ovate colleters (in whole flower 3–10). Corolla subhypocrateriform; tube yellow and violet or dark-red, ventricose in the middle, outside glabrous at the base and minutely pubescent at the apex, inside at the apex just below the appendages with a narrow pilose zone; lobes yellow with a brownish-violet tinge, in the bud overlapping to the left, elliptic, rounded, entire, ciliate, minutely pubescent outside, with an alternating set of appendages inside. Appendages bright yellow, in a set of 1–3, shortly truncate or filiform, glabrous or pubescent, plus 2–4 long filiform glabrous ones. Stamens exserted, but covered by the appendages; filaments short, filiform, glabrous outside, inside hispid towards the apex and with a triangular swelling on the base of the connective; anthers introrse, sagittate at the base, apiculate and pilose at the apex, otherwise glabrous, fertile for ⅓ of the length just below the apices. Pistil glabrous or sometimes with a few papillae; ovary superior, subglobose; carpels two, free, rounded; disk none; style split at the base; clavuncula with a thin ring at the base, almost cylindrical, covered by the anthers and inside with the triangular swellings of the connectives adnate to the clavuncula; stigma minute, bilobed. Ovules numerous,

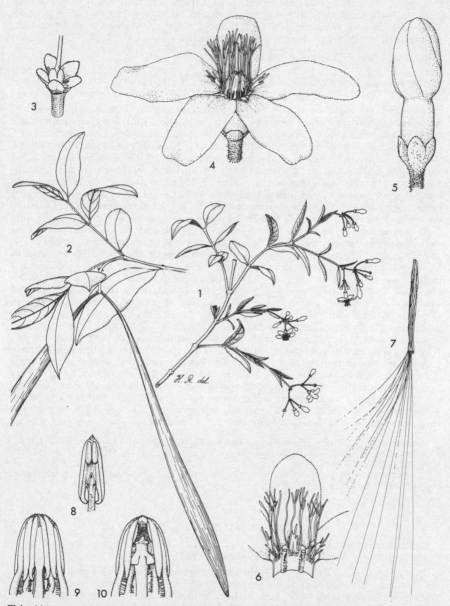

Tab. 114. PLEIOCERAS ORIENTALE. 1, flowering branch (× ⅓), from *Vollesen* MRC 4783; 2, fruiting branch (× ½), from *Simão* 724; 3, calyx with pistil (× 2½); 4, flower (× 2½); 5, flower bud (× 2½), 3–5 from *Vollesen* MRC 4783; 6, part of corolla with appendages (× 7), from *Vollesen* MRC 4222; 7, seed (× 1½), from *Simão* 724; 8, anther inside (× 7); 9, anthers outside (× 7); 10, anthers with pistil-head (× 7), 8–10 from *Vollesen* MRC 4783. Reprinted from Bot. Tidsskr. **75**: t. 2 (1980).

in several rows. Follicles 2, spreading, almost free, pendulous, slender, adaxially dehiscent, glabrous; wall thinly coriaceous, striate, grooved when dry, glabrous outside. Seeds numerous, in two or more rows, linear to very narrowly oblong, finely grooved, glabrous, with a ± dense tuft of retrorse hairs at the apices, embryo large, surrounded by the very scanty whitish endosperm.

A genus of 5 species in tropical Africa.

Pleioceras orientale Vollesen in Bot. Tidsskr. **75**: 59, tab. 2 (1980). — Barink in Meded. Landb. Wag. **83**–7: 37, fig. 4, map 4 (1984). TAB. **114**. Type from Tanzania.

Tree, 5–8 m. high; bark pale brown. Branches pale to dark brown when dry; branchlets densely pilose. Leaves petiolate; petiole 3–5 mm. long, densely pilose; lamina narrowly ovate or obovate, 2·6–3·5 times as long as wide, 3·5–14·5 × 1–5·5 cm., subcoriaceous when dry, puberulous above, beneath tomentose especially on the main veins. Inflorescences several-flowered, 4·5–6·5 × 4–8·5 cm. Bracts 2·2–3 mm. long, pubescent or glabrous on both surfaces. Peduncle 1–2 cm. long, tomentose, rarely subglabrous; pedicels 7–15 mm. long, tomentose, rarely subglabrous. Sepals 1–1·7 times as long as wide, 2–3 × 1–2 mm., sparsely pubescent outside, glabrous inside with or without a colleter at the base near the edge. Colleters (in whole flower 3–5) large, triangular, 2·5 times as long as wide, 1 × 0·4 mm. Corolla 6–7 times as long as the calyx, 13·5–20 mm. long; tube dark violet, 3–4 times as long as the calyx, 6–9 mm. long, at the ventricose part 2·5–4 mm. in diam., inside below the pilose zone glabrous; lobes yellow, brownish-violet at the apex, 1·2–1·8 times as long as the tube, 1·8–2 times as long as wide, 7·5–11 × 4–6 mm., minutely pubescent outside, papillose inside. The set of appendages composed of 7 (the 1st and the 7th small, the 2nd and 3rd broom-like, the 4th, the central, much shorter, truncate and bearing three narrowly oblong lobes on the edges and in between the edges, the 5th and 6th again broom-like; the small ones oblong, 0·5–1·5 × 0·5 mm., ciliate, glabrous; the broom-like appendages long, filiform, 7–9 mm. long, glabrous, branched from 5–6 mm. above the base into 1–3 branches; the central one oblong, 2–3 × 0·3 mm., branched from 1·3–1·8 mm. above the base into 3 branches, sparsely pubescent). Stamens inserted 1·5–1·6 mm. below the mouth of the corolla; filaments 2·2–3 × 0·2 mm.; anthers 2·5–3·2 × 1 mm. Pistil 8–9 mm. long; ovary 1–1·5 × 0·6–0·8 × 0·5–1 mm., glabrous; style 6–8 × 0·2–0·3 mm., glabrous; clavuncula 0·8–1 × 0·9–1 mm.; stigma 0·2–0·3 mm. long. Ovules 120–150 in each carpel. Fruit dark green; follicles 20–30 cm. long, 5–7 mm. in diam. Seeds 10–18 × 1 × 0·2 mm., greenish, with a dense, 35 mm. long tuft of hairs at the apex.

Mozambique. Z: 20 km. from Mopeia to Massingire, fr. 30.vii.1942, *Torre* 4463 (LISC). MS: 25 km. from Lacerdónia, N. of new railway 200 m., fl. 6.xii.1971, *Müller & Pope* 1924 (C; K; LISC; SRGH; WAG).
Lowland dry evergreen forest, deciduous coastal forest on sand and deciduous coastal thicket on sand. Alt. 200–400 m.

25. FUNTUMIA Stapf

Funtumia Stapf in Proc. Linn. Soc. **1899**: 2 (1899); in Hooker, Ic. Pl. **27**: t. 2694–2695 (1901).

Evergreen trees or shrubs, with white sticky latex; bark smooth, sometimes with a few lenticels, greenish-brown to grey; wood of low density, soft; branches terete, sometimes lenticellate, very dark brown; branchlets smooth, terete or laterally compressed with a longitudinal groove below the ocrea. Leaves opposite, petiolate, those of a pair connate into a short ocrea, with many small colleters in two or three rows in the axils; lamina ovate, elliptic or oblong, decurrent into the petiole, acuminate at the apex, entire, glabrous above; often domatia in the axils of the secondary veins; margin both undulate and revolute. Inflorescences congested, terminal and axillary, cymose, much shorter than the leaves; peduncle short; bracts ovate or elliptic, often with small colleters in the axils. Flowers actinomorphic, fleshy, fragrant. Sepals free, thick, ovate or nearly so, obtuse or subacute at the apex, often membranous and minutely ciliate at the margin, inside with a single row of colleters at the base. Corolla tube ventricose at the middle, inside thickened at the throat, densely hirto-pubescent from the insertion of the stamens to the level of the apex of the ovary, with an indumentum which gradually becomes thinner and shorter; lobes overlapping in bud to the right, often auriculate at the left, entire, recurved. Stamens included; filaments very short or absent, ventrally densely hirto-pubescent, dorsally glabrous; anthers narrowly triangular with a very short point at the acuminate apex, sagittate at the base, glabrous, 2-celled, introrse; connective dorsally appressed-pubescent, ventrally at the base stiffly coherent with the clavuncula. Ovary composed of 2 almost free carpels which are united by the base of the style; placenta adaxial, 2-lobed; style not split at the base, with two longitudinal grooves, thickened below the clavuncula; clavuncula grading into the stigma, together ovoid; stigma conical. Disk 5-lobed; lobes truncate or acute, minutely toothed at the apex. Ovules pendulous, 200–350 in each carpel. Fruit of two follicles which are connate at the base, green and glossy when young, turning grey-brown and woody when maturing, striate; follicles adaxially flattened and there dehiscent, abaxially convex, sometimes slightly curved; wall woody, thick, outside grey-brown smooth and yellowish-brown inside. Seed slender, fusiform, with the apex rostrate; rostrum at least above the middle with long straight hairs enveloping the seed within the fruit; testa rugose; endosperm white, surrounding the embryo; embryo white, straight, about 0·9 times as long as the seed.

A genus of 2 species in tropical Africa from Senegal to Tanzania and Zimbabwe.

Funtumia africana (Benth.) Stapf in Hooker, Ic. Pl. **27**: t. 2696–2697 (1901). — Stapf in F.T.A. **4**, 1: 190 (1902). — H. Huber in F.W.T.A., ed. 2, **2**: 74 (1963). — Zwetsloot in Meded. Landb. Wag. **81**–16: 16, fig. 3, phot. 4, 6–8, map 2 (1981). TAB. **115**. Type from Fernando Po.
   Kickxia africana Benth in Hooker, Ic. Pl. **13**: t. 1276 (1879). Type as above.
   Kickxia latifolia Stapf in Kew Bull. **1898**: 307 (1898). Type from Zaire.
   Funtumia latifolia (Stapf) Stapf in Proc. Linn. Soc. **1899**: 2 (1899). Type as for the species above.
   Kickxia scheffleri K. Schum. in Notizbl. Bot. Gart. Berlin **3**: 81 (1900). Type from Tanzania.
   Kickxia zenkeri K. Schum., loc. cit. Type from Cameroon.
   Kickxia congolana De Wild. in Rev. Cult. Col. **7**: 745 (1900). Type from Zaire.
   Kickxia gilletii De Wild., loc. cit. Type from Zaire.
   Funtumia scheffleri (K. Schum.) Jumelle, Pl. Caoutch. 381 (1903). Type as for *Kickxia scheffleri*.
   Funtumia zenkeri (K. Schum.) Jumelle, loc, cit. Type as for *Kickxia zenkeri*.
   Funtumia congolana (De Wild.) Jumelle, loc. cit. Type as for *Kickxia congolana*.
   Funtumia gilletii (De Wild.) Jumelle, loc. cit. Type as for *Kickxia gilletii*.

Tree or shrub, up to 30 m. high. Trunk up to 50 cm. in diam.; branchlets glabrous or minutely pubescent. Leaves petiolate; petiole 0·3–1·5 cm. long, glabrous or minutely pubescent; lamina subcoriaceous to coriaceous, 1·5–4 times as long as wide, 5–32 × 1·7–17 cm., often puberulous on the midrib; secondary veins 6–13, mostly parallel; domatia consisting of a tuft of straight hairs, varying in density, sometimes absent; margin slightly undulate and somewhat revolute. Inflorescence 3–40-flowered, 2–2·5 × 1·5–4·5 × 1·5–4·5 cm.; pedicels 3–15 mm. long, glabrous or puberulous; bracts obtuse or acute, 0·9–2 mm. long with colleters in the axils. Mature buds cylindrical to slightly conical, 8·5–22 mm. long, obtuse to subacute. Sepals triangular to broadly ovate, 1·5–4 × 1·8–3 mm., glabrous or puberulous outside; colleters, 5–50 varying in both shape and size, even in a single flower. Corolla tube very pale green to creamy, almost cylindrical, ventricose at $\frac{2}{5}$–$\frac{4}{5}$ from the base, 1·7–5 times as long as the calyx, 5·8–10 × 2·5–5 mm., glabrous to puberulous outside; lobes creamy, recurved, 0·5–1·6 times as long as the tube, obliquely ovate to narrowly oblong, 1–5·5 times as long as wide, 5–15 × 2–5 mm., obtuse or acute, glabrous to puberulous on both sides, sometimes pubescent inside near the base. Stamens 1·1–4·5 × 0·6–1·5 mm., inserted at $\frac{2}{5}$–$\frac{4}{5}$ from the base of the tube; anthers 1–4·5 × 0·6–1·5 mm. with a 1·1–2·2 mm. long fertile portion; connective dorsally appressed pubescent. Pistil 4–7 mm. long; ovary almost cylindrical to subglobose. 1·2–2 × 0·9–2 mm., pubescent at the apex, style 1·2–3 × 0·2–0·35 mm., glabrous or with a few erect hairs; clavuncula and stigma 0·9–2·4 × 0·3–1·1 mm. Disk 5-lobed; lobes 0·4–1·6 mm. long, shorter than or rarely as long as the ovary. Fruit a follicle 8·4–32 cm. long, almost fusiform, 1·4–5·1 cm. in circumference and with the flat adaxial side 0·6–3·6 cm. wide. Seed 3·5–7·5 × 0·2–0·6 cm.; rostrum 1·6–6 cm. long with 3·2–9 cm. long hairs.

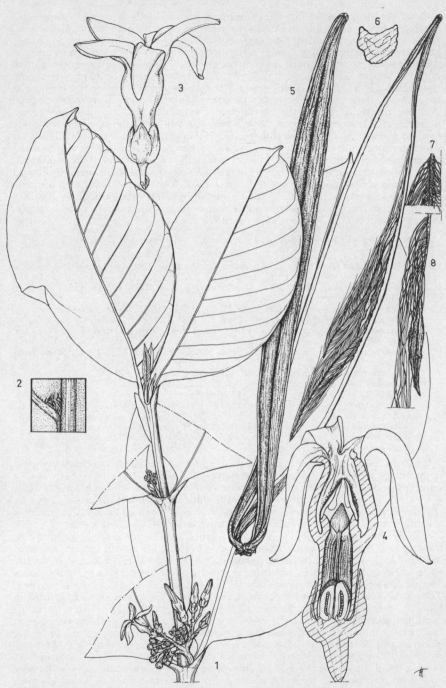

Tab. 115. FUNTUMIA AFRICANA. 1, flowering branch ($\times \frac{2}{3}$), 2, domatium in the axil of the secondary vein ($\times 4$), 3, flower ($\times 2$), 4, longitudinal section of the flower ($\times 4$), 5, opened fruit ($\times \frac{2}{3}$), 6, transverse section of the fruit ($\times \frac{2}{3}$), 7, detail of the haired beak ($\times 2$), 8, detail of seed ($\times 2$), 1–4 from *Zwetsloot* 28, 5–8 from *Zwetsloot* 21.

Zimbabwe. E: Chimanimani (Melsetter), bank of Lusitu R., near junction with Haroni R. fl. 25.xi.1955, *Drummond* 5024 (K; LISC; LMA; S; SRGH). **Malawi.** S: Thyolo Distr., Cholomwani, 600 m., *Topham* 699 (FHO). **Mozambique.** N: Chomba, 2.xi.1959, *Gomes e Sousa* 4516 (COI; K; LMA; LMU; MO; SRGH). MS: Sofala Prov., between Chiniziua and Macalaua Rs., fl. 16.iv.1957, *Gomes e Sousa* 4363 (COI; FHO; FI; K; LISC; LMA; M; P; SRGH).

Also in tropical Africa from Senegal to Tanzania. Moist light or secondary forest, in savanna area of riverine forest. Alt. 0–1300 m.

26. MASCARENHASIA A. DC.

Mascarenhasia A. DC. in DC., Prodr. **8**: 487 (1844). — Pichon in Mém. Inst. Sci. Madag., Sér. B, **2**: 76 (1949).

Shrubs or small trees with milky latex. Spines and tendrils absent. Leaves opposite, exstipulate, without domatia, with petiolar glands. Flowers often large and conspicuous, solitary or in few-flowered fasciculate cymes, terminal. Calyx lobes imbricate, free to base, with many glandular scales inside at the base. Corolla hypocrateriform; tube divided into two distinct regions of varying shape and proportions; lobes 5, with induplicate aestivation, straight in bud or twisted to the right. Stamens 5, sessile, inserted at the base of the upper region of the corolla tube; anthers conniving in a cone round the gynoecium, polliniferous in upper part only, joined to the clavuncle by a retinacle. Receptacular disc cupular or represented by separate oppositipetalous scales, these sometimes united in pairs. Ovary of 2 free multiovulate carpels, glabrous or hairy; style pubescent; clavuncle swollen; stigma pointed or flared at apex, not bifid. Fruit of 2 follicular mericarps. Seeds linear-compressed with an apical coma; endosperm absent.

A genus of 10 species, 9 endemic to Madagascar and one distributed from Madagascar and the Comores to tropical Africa.

Mascarenhasia arborescens A. DC. in DC., Prodr. **8**: 488 (1844). — Pichon in Mém. Inst. Sci. Madag., Sér. B, **2**: 81 (1949). — R. B. Drumm. in Kirkia **10**: 269 (1975). TAB. **116**. Type from Madagascar.
   Mascarenhasia variegata Britten & Rendle in Trans. Linn. Soc., Ser. 2, **4**: 26, t. 6 figs. 1–3 (1894). — Stapf in F.T.A. **4**, 1: 193 (1902). Type: Malawi, Mt. Mulanje, 1800 m., x.1891, *Whyte* 108 (K, holotype; BM, isotype).
   Mascarenhasia fischeri K. Schum. in Engl., Pflanzenw. Ost-Afr. **C**: 318 (1895). Type from Tanzania.
   Mascarenhasia elastica K. Schum. in Notizbl. Bot. Gart. Berl. **2**: 270 (1899). — Stapf, op. cit.: 194 (1902). — Dale & Greenway, Kenya Trees & Shrubs: 47 (1961). Type from Tanzania.
   Lanugia latifolia N.E. Br. in Torreya **27**: 52 (1927). Type a cultivated plant grown in S. America from "seeds from Mozambique" (K, holotype).
   Lanugia variegata (Britten & Rendle) N.E. Br., tom. cit.: 53 (1927). Type as for *Mascarenhasia variegata*.

Shrub or small tree 1·5–8(12) m. high; young shoots glabrous or brown appressed-pubescent and soon glabrescent; bark of twigs grey, rough with lenticels. Leaves glabrous or petiole and proximal part of lower lamina surface brown appressed-pubescent and soon glabrescent; petiole 4–8 mm. long; leaves drying discolorous, with upper surface blackish green or dark brown and lower surface paler, grey or brown. Leaf lamina 5–16·5 × 1·7–6 cm., obovate-oblong, oblong or elliptic, the apex cuspidate or acuminate to a round-tipped acumen, the base tapered; main lateral veins 7–13-paired, looping into a continuous marginal vein, level and inconspicuous on upper surface, slightly raised and quite conspicuous on lower surface; tertiary reticulum not scalariform; midrib impressed above, prominent below. Inflorescences few- or several-flowered cymes or fascicles, these terminal but often appearing lateral by the overtopping growth of axillary shoots; pedicels 2–7 mm. long, glabrous or pubescent; flowers white with yellow tube, the corolla leathery in texture. Calyx 2·5–4 mm. long, lobes ovate-acute, thinly appressed-pilose on both surfaces, ciliate; a ring of numerous tiny scales present, adnate to the base of the calyx on its adaxial side. Corolla lobes ovate-acute to ovate-caudate, 5–10 mm. long, densely hairy on both surfaces and on the margin, the indumentum consisting of turgid cells longest towards the mouth of the corolla tube; corolla tube 8–12 mm. long, of 2 ± equal parts separated by a constriction and with a very small apical

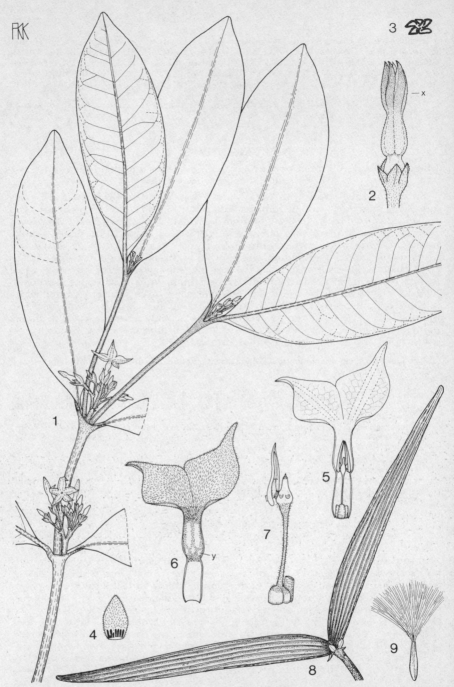

Tab. 116. MASCARENHASIA ARBORESCENS. 1, habit (× ⅔), from *Torre* 654; 2, flower bud (× 2); 3, cross-section through bud at level "x" showing induplicate aestivation (× 2), 2–3, from *Rail* 1/55; 4, calyx lobe, adaxial view, showing the tiny scales (× 4); 5, part of corolla opened out showing venation of corolla lobes and position of androecium, gynoecium and disc scales; indumentum omitted (× 2); 6, as 5, corolla alone, showing indumentum; the hairless areas "y" are points of stamen insertion (× 2); 7, one stamen, the gynoecium, and one disc scale (× 4), 4–7 from *Mendonça* 2478; 8, fruit (× ⅔), from *Barbosa* 2115; 9, seed (× ⅔), from *Andrada* 1911.

opening; outer surface glabrous or pubescent especially in the upper half, inner surface variably pilose in the upper half, especially behind the anthers, ± glabrous below the constriction. Stamens sessile, fixed by their lower abaxial surface to the corolla tube just above the median constriction; anthers sagittate, c. 3·5 mm. long; pollen sacs confined to the upper half of the anther, the lower half consisting of two stiff sterile tails flanking a median brush-like retinacle which is firmly adnate to the clavuncle of the gynoecium. Disc bearing 4–5 stout oblong scales c. 1·6 mm. high. Ovary of two completely separate compressed-ovoid carpels c. 1 mm. high, pubescent at the apex. Style densely pilose, widened at apex at level of corolla tube constriction into the clavuncle, then narrowed again into the cylindrical, glabrous, apparently undivided stigma. Fruit of two straight divaricate follicles 5–17·5 cm. long, with rough grey surface. Seeds ∞, 11–15 mm. long, linear, compressed, with apical coma of golden-brown hairs 16–27 mm. long.

Zimbabwe. E: Chimanimani Mts., foot of Mt. Peza, 1050 m., fr. 12.x.1950, *Wild* 3577 (K; LISC; PRE; SRGH). **Malawi.** S: Likabula Gorge, 840 m., fr. 20.vi.1946, *Brass* 16369 (K; PRE; SRGH). **Mozambique.** N: R. Malema near Malema village, fl. & fr. 28.x.1954, *Gomes e Sousa* 4270 (COI; K; LISC; PRE; SRGH). Z: Alto Molócuè, Gilé to Mamala, near R. Nàpomé, fr. 31.viii.1949, *Andrada* 1911 (COI; LISC). MS: Sofala Prov. (Beira Distr.), R. Chiniziua, fl. & fr. immat. 13.iv.1957, *Gomes e Sousa* 4358 (COI; K; LISC; PRE).

Occurring in Madagascar, the Comores, Kenya, Tanzania, Zaire and Togo (cultivated in Cameroon). In riverine forest.

The very numerous specimens of *M. arborescens* seen from the F.Z. area all give a height for this plant within the range 1·5–8(12) m. except for *Müller* 1121 from the Chimanimani Mts. which describes it as a tree of 21 m. with trunk strongly fluted at the base. Such a difference in stature is puzzling, and one wonders if a mistake could have been made in the collection of the latter specimen; if not, it would be interesting to investigate the situation in which the species can attain this size.

Markgraf, in Adansonia **12**: 588 (1973), has divided *M. arborescens* into five varieties of which only var. *comorensis* Markgraf is said to occur naturally on the African mainland. This variety is characterised by leaves 9–12 × 3–5 cm., and flowers with lower part of corolla tube 3 mm., upper part of tube 3 mm. and corolla lobes 7 mm. long. It is clear that these measurements do not correspond with those of specimens from the F.Z. area; therefore I cannot recognise var. *comorensis*.

27. ALAFIA Thouars

Alafia Thouars, Gen. Nov. Madag.: 11 (1809). — Pichon in Bull. Jard. Bot. Brux. **24**: 131 (1954).

Scandent shrubs and lianes with milky latex. Spines and tendrils absent. Leaves opposite, with a pair of triangular stipules sometimes joined into a single intrapetiolar scale, without acarodomatia or petiolar glands. Flowers borne in terminal umbellate cymes. Calyx lobes imbricate, free to base, each overlapped margin with one or a pair of delicate scales inside at the base. Corolla hypocrateriform; tube ± cylindrical, slightly wider at the middle, glabrous on outer surface except for faint alternipetalous lines of puberulence, ± glabrous within except for a circle of long stiff downwards-pointing hairs below the level of stamen-insertion; corolla lobes 5, contorted, overlapping to the right, indumentum variable. Stamens 5, sessile, inserted at about the middle of the corolla tube and with their tips reaching to the mouth of the tube; anthers conniving in a cone round the gynoecium, polliniferous in upper part only, joined to the clavuncle by a retinacle; dorsal surface of anthers often pubescent. Receptacular disc absent. Ovary of 2 free multiovulate carpels pubescent at apex; style glabrous, gradually dilated towards the clavuncle, contracted and then widened into the capitate stigma. Fruit of 2 follicular mericarps. Seeds linear-compressed with an apical coma; endosperm absent.

A genus of c. 30 species, 11 native to Madagascar and the rest to tropical Africa.

1. Leaf lamina with 8–11 pairs of main lateral nerves; corolla lobes much longer than corolla tube, pubescent on upper surface - - - - - - - - 3. *microstylis*
 - Leaf lamina with 4–7 pairs of main lateral nerves; corolla lobes shorter than to slightly exceeding corolla tube, glabrous on upper surface - - - - - - - 2
2. Inflorescence a very dense head of numerous flowers; corolla red on outer surface; follicles up to 70 cm. long; leaf lamina puckered along the veins - - - - 4. *orientalis*
 - Inflorescence with up to 17 flowers in an umbellate cyme, not densely crowded; corolla

uniformly white or cream-coloured; follicles up to 36 cm. long; leaf lamina not puckered
     along the veins - - - - - - - - - - - - - - - - 3
3. Leaf-lamina scarcely apiculate; corolla lobes 2–3·5 mm. wide -   -   - 1. *zambesiaca*
– Leaf-lamina with distinct slender apiculus; corolla lobes 4–8 mm. wide
                                                      2. *caudata* subsp. *latiloba*

1. **Alafia zambesiaca** Kupicha in Bull. Jard. Bot. Nat. Belg. **51**: 153, fig. 1, 4 (1981). TAB.
    117. Type: Zambia, Kasama Distr., 5 km. E. of Kasama, fl. 5.xi.1960, *Robinson* 4041 (K,
    holotype; PRE, SRGH, isotypes).
        Alafia caudata sensu F. White, F.F.N.R.: 346, fig. 62D, F (1962). — Fanshawe, Check
    List Woody Pl. Zambia: 3 (1973). — R. B. Drumm. in Kirkia **10**: 269 (1975) non Stapf
    (1894).

Scandent shrub or liane 1·5–15 m. high, sometimes forming thickets. Young stems
pubescent, soon glabrescent; bark of older twigs grey, faintly longitudinally
wrinkled and with inconspicuous pale dot-like lenticels. Leaves thinly coriaceous,
undulate, drying almost concolorous brown or green but paler beneath; petioles 1–2
mm. long, pubescent but soon glabrescent. Leaf-lamina 18–60 × 9–28 mm., ovate to
lanceolate, glabrous; apex obtuse or acute, usually slightly apiculate; base cuneate;
margin faintly revolute; main lateral nerves 5–7-paired; midrib and lateral nerves
impressed on upper surface, prominent and fairly conspicuous on lower surface.
Inflorescences 4–17-flowered, occasionally lateral as well as terminal; peduncles 0–7
long, pedicels 3–5 mm.; flowers white or creamy white, heavily scented. Calyx
1·5–2·5 mm. long; sepals free, lanceolate, ciliate, otherwise glabrous, their scales
paired or rarely single or absent. Corolla tube 6·75–9·5 mm. long; corolla lobes
6·5–10·5 × 2–3·5 mm., 2·5–4·2 times as long as broad, ± linear, glabrous except for
the ciliate margin and triangle of papillose indumentum at the mouth of the corolla
tube. Stamens inserted at 3·2–4·4 mm. from the base of the corolla tube (at ±½ of its
height); anthers 3–3·5 mm. long, dorsal surface pubescent or glabrous. Ovary c. 1
mm. long. Follicles 17–30 cm. long, dark grey, smooth, curved, slightly torulose.
Seeds 10–17 mm. long; coma golden-brown, 25–44 mm. long.

Zambia. N: Mbala (Abercorn) Distr., 1·6 km. E. of Mpulungu, fl. 17.xi.1952, *Angus* 779
(BR; FHO; K). W: Chingola, fl. 6.xi.1953, *Fanshawe* 482 (BR; K). C: 1·6 km. S. of Chimbala,
fr. 13.iv.1958, *Angus* 1898 (BR; FHO; K; PRE; SRGH). S: Mazabuka Distr., 16 km. from
Choma to Pemba, fr. 30.i.1960, *White* 6633 (FHO; K). **Zimbabwe.** N: Chipuriro (Sipolilo)
Distr., E. of Mt. Chiruwa, c. 800 m., fr. 30.i.1966, *Müller* 314 (SRGH). N/W: Sebungwe
Distr., near Dett R., fl. 7.ix.1956, *Lovemore* 470 (K; LISC; SRGH). C: Kadoma (Gatooma)
Distr., 1·6 km. N.W. of Kadoma, c. 1160 m., fr. 3.vi.1968, *Rushworth* 1163 (SRGH). E:
Chipinge (Chipinga) Distr., 27 km. S. of Chirinda, near Mwangazi area, fr. vii.1964, *Goldsmith*
36/64 (BR; K; LISC; PRE; SRGH).
    Also occurring in Zaire (Shaba) and Tanzania. In savanna woodland, in thickets, on
termitaria, beside streams among rocks.

2. **Alafia caudata** Stapf in Kew Bull. **1894**: 123 (1894). Type from Angola.

Subsp. **latiloba** Kupicha in Bull. Jard. Bot. Nat. Belg. **51**: 161, fig. 4 (1981). Type from
    Tanzania.

Climbing shrub or liane 1–25 m. tall. Young stems glabrous; bark of older twigs
smooth, grey or brown, with dot-like lenticels. Leaves thinly coriaceous, glabrous,
with the tip depressed so that the lamina folds when pressed flat, drying dark
brownish green above and paler below; petioles c. 3 mm. long. Leaf lamina
4·5–6·75 × 1·8–3·5 cm., elliptic with apex cuspidate to a distinct slender apiculus
("drip-tip"), base attenuate; margin flat, not revolute; main lateral nerves 4–5-
paired; midrib and lateral nerves faintly impressed above, faintly raised below.
Inflorescences 5–13-flowered, with peduncle 5–13 mm. long and pedicels 6–10 mm.
long. Flowers white to yellowish, scented. Calyx 1·5–2 mm. long; sepals free,
lanceolate, ciliate, otherwise glabrous; scales paired. Corolla tube 7·5–9 mm. long;
corolla lobes 6–13 × 4–8 mm., 1·3–1·9 times as long as broad, ± rounded-triangular,
glabrous except for the ciliate margin and triangle of papillose indumentum at the
mouth of the corolla tube. Stamens inserted at 4·2–5·5 mm. from the base of the
corolla tube (at ½–⅔ of its height); anthers 3–3·5 mm. long, pubescent on dorsal
surface. Ovary c. 1 mm. long. Follicles c. 36 cm. long, slender, smooth, grey, slightly
curved, faintly torulose. Seeds c. 16 mm. long; coma golden brown, 35–40 mm. long.

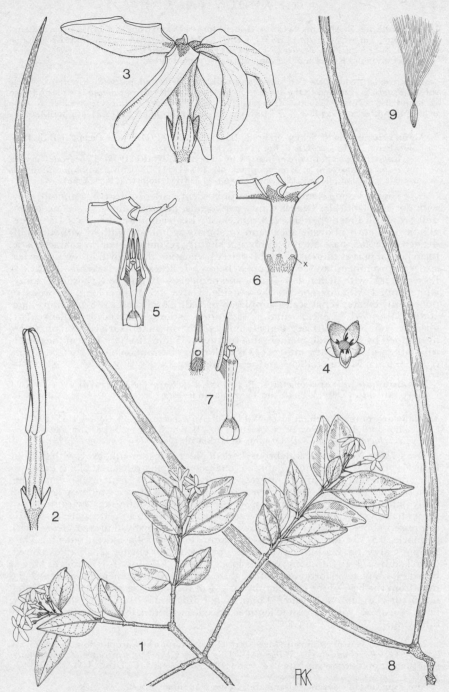

Tab. 117. ALAFIA ZAMBESIACA. 1, habit ($\times\frac{2}{3}$), from *Thomson* 501; 2, flower bud ($\times 4$); 3, flower ($\times 4$); 4, calyx, seen from above with rest of flower removed, showing the scales (black) on each overlapped sepal margin ($\times 6$); 5, half-flower, without corolla indumentum ($\times 4$); 6, the same as 5, corolla alone, showing indumentum, "x" representing position of stamen insertion ($\times 4$); 7, gynoecium and a single stamen in side and ventral views ($\times 6$), 2–7 from *Robinson* 4041; 8, fruit ($\times\frac{2}{3}$); 9, seed ($\times\frac{2}{3}$), 8–9 from *Bingham* 685.

Mozambique. Z: Maganja da Costa, Gobene forest, 48 km. from Vila da Maganja, c. 20 m., fl. 12.ii.1966, *Torre & Correia* 14574 (C; LISC; LMA; WAG). MS: 25 km. from Lacerdónia, 200 m., fl. buds 6.xii.1971, *Müller & Pope* 1922 (K; LISC; SRGH).
Also occurring in Kenya and Tanzania. In dense mixed forest.

Alafia caudata comprises two subspecies with completely allopatric distributions. While subsp. *latiloba* is confined to the Indian Ocean coastal belt, subsp. *caudata* is widespread in Zaire and also occurs in Gabon and Angola. The typical subspecies differs from subsp. *latiloba* in having shorter corolla tubes (5–7·5 mm.) and narrower corolla lobes (2.25–4(5) mm.).

3. **Alafia microstylis** K. Schum. in Engl., Bot. Jahrb. **23**: 230 (1896). — Pichon in Bull. Jard. Bot. Brux. **24**: 188, t. 4E & F, fig. 17 (1954). Syntypes from Uganda.
    Alafia clusioides S. Moore in Journ. Linn. Soc., Bot. **37**: 181 (1905). Type from Uganda.
    Alafia swynnertonii S. Moore, op. cit. **40**: 141 (1911). Type: Mozambique, Madanda forests, c. 120 m., fl. 5.xii.1906, *Swynnerton* 1178 (BM, holotype; K, isotype).

Large liane. Young stems glabrous or pubescent, bark of older twigs smooth, grey, with dot-like lenticels. Leaves thinly coriaceous, glabrous, drying $\pm$ concolorous dull brownish green; petioles 2–8 mm. long. Leaf-lamina c. 4·5–7·5 × 1·5–3 cm., elliptic or slightly obovate, apex acute or shortly acuminate, with or without an ill-defined apiculus, base attenuate; margin slightly revolute; venation characteristic: main lateral nerves numerous (8–11-paired), venation inconspicuous on upper leaf surface, prominent and conspicuous below. Inflorescences several- to many-flowered, lax, with peduncle 1–7 mm. and pedicels 4–10 mm. long; corolla creamy-white tinged with maroon on the outside. Calyx 1·5–3 mm. long, glabrous or pubescent, ciliate; sepal scales simple or paired. Corolla tube 4·5–6·5 mm. long; corolla lobes 9–13 × 3·5–5·5 mm., 2–3·5 times as long as broad, $\pm$ oblanceolate, glabrous on outer surface, densely pubescent on inner surface except where overlapped by next petal. Stamens inserted at 2–2·5 mm. from the base of the corolla tube (at $\pm\frac{1}{2}$ of its height); anthers c. 3 mm. long, pubescent on dorsal surface. Ovary c. 1 mm. long. Mature follicles and seeds unknown.

Mozambique. MS: Amatong forest, fl. xii.1963, *Pole Evans* 6683 (K; PRE).
Also occurring in Zaire, Uganda and Tanzania. In forest.

4. **Alafia orientalis** K. Schum. ex De Wild., Not. Apocinac. Latic. Fl. Congo, **1**: 15 (1903). — Pichon in Bull. Jard. Bot. Brux. **24**: 206, t. 6A, B, fig. 20 (1954). Type from Tanzania.
    A. schumannii sensu R. B. Drumm. in Kirkia **10**: 269 (1975), non Stapf (1894).

Large liane. Young stems glabrous, bark of older twigs smooth, grey, with dot-like lenticels. Leaves coriaceous, glabrous, characteristically puckered along the veins when dry, drying dark brown or green above, paler below; petioles 3–7 mm. long. Leaf lamina 4·5–8 × 2·2–4·4 cm., ovate or ovate-elliptic, apex cuspidate to a short or fairly long, round-tipped apiculus, base rounded or cuneate; margin flat or somewhat revolute; main lateral nerves 4–5-paired; midrib and lateral nerves faintly impressed above, faintly raised below. Inflorescences many-flowered, dense, with peduncle 1·5–10 mm. and pedicels 3–5 mm. long; corolla red on outside, white within. Calyx 1·5–2·5 mm. long, lobes pubescent and ciliate; sepal scales paired. Corolla tube 5·5–10 mm. long; corolla lobes 4·5–8 × 3–5 mm., 1–1·7 times as long as broad, $\pm$ obovate, glabrous except for the ciliate margin. Stamens inserted at 2–3·5 mm. from the base of the corolla tube (at $\frac{1}{3}-\frac{1}{2}$ of its height); anthers 4–4·5 mm. long, dorsal surface glabrous. Ovary c. 1 mm. long. Follicles 25–70 cm. long, smooth, grey, faintly torulose, without visible lenticels. Seeds c. 18 mm. long; coma golden, c. 50 mm. long.

Zimbabwe. E: Chimanimani (Melsetter) Distr., Haroni/Makurupini forest, 400 m., fr. 3.xii.1964, *Wild, Goldsmith & Müller* 6619 (BR; K; SRGH). **Mozambique.** Z: Massingire, near summit of Morrumbala Mt., fl. & fr. 6.v.1942, *Torre* 4504 (C; LISC; LMA; MO; PRE; WAG).
Also occurring in Tanzania (Usambara). In dense montane forest.

28. ONCINOTIS Benth.

Oncinotis Benth. in Hook., Niger Fl.: 451 (1849). — Pichon in Mém. Mus. Nation. Hist. Nat. Paris, Sér. B, **1**: 1–143 (1950); in Bull. Jard. Bot. Brux. **24**: 9–36 (1954).

Scandent shrubs and tall lianes with milky latex. Spines, tendrils and stipules absent. Glands present at or near apex of petiole. Domatia usually present on lower leaf surface. Flowers in terminal or axillary paniculate cymes, often greenish and inconspicuous. Calyx lobes imbricate, free almost to base. Corolla hypocrateriform; tube short, widest at the middle, hairy inside and outside, bearing 5 alternipetalous corona scales at the mouth; corolla lobes 5, contorted, overlapping to the right, spreading to reflexed, ± equalling tube. Stamens inserted almost at base of corolla tube; anthers large, conniving in a cone round the gynoecium, fertile in the upper part only, joined to the clavuncle by a retinacle (a linear appendage produced from the connective at the base of the anther). Receptacular disc cupular, with 5 oppositipetalous lobes. Ovary of 2 multiovulate carpels connate at the extreme base only; style very short; clavuncula consisting of a slightly 5-winged upper and a cylindrical lower portion, to the latter of which the 5 retinacles adhere; stigma bilobed. Fruit of 2 follicular mericarps. Seeds oblong with an apical coma; endosperm in a thin layer completely surrounding the embryo.

A genus of 7 species, native to Africa and Madagascar.

Oncinotis tenuiloba Stapf in Kew Bull. **1898**: 307 (1898). Type from Zaire. — Stapf in F.C. **4**, 1: 512 (1907). — Pichon in Bull. Jard. Bot. Brux. **24**: 24, t. 2A, fig. 2 (1954). — Codd in Fl. Southern Afr. **26**: 288 (1963). TAB. **118**. Lectotype (of Pichon) from S. Africa (Natal).

Motandra erlangeri K. Schum. in Engl., Bot. Jahrb. **33**: 318 (1903). Type from Ethiopia.

Oncinotis inandensis J. M. Wood & Evans in Journ. Bot. Lond. **37**: 254, pl. 61 (1899).

Oncinotis natalensis Stapf in Kew Bull. **1907**: 52 (1907). Syntypes as for *Oncinotis inandensis*.

Oncinotis chirindica S. Moore in Journ. Linn. Soc., Bot. **4**: 141 (1911). Type: Zimbabwe, Chirinda Forest, 1130–1220 m., fl. & fr. 8.x.1906, *Swynnerton* 87 (BM, holotype; K, SRGH, isotypes).

A large, strong liane reaching 30 m. or more in the crowns of trees, or a scrambling shrub, with light ochraceous thick corky bark and plentiful milky latex. Stems flexuose, at first usually reddish-brown or greyish puberulous with a mixture of simple and branched hairs, later glabrescent with conspicuous whitish dot-like lenticels. Leaves membranous, drying ± concolorous green; petiole 5–10 mm. long, puberulous; 2 minute oblong or triangular glands present on upper leaf surface at proximal end of midrib; lamina 6·2–13 × 2–5 cm., narrowly obovate or obovate-oblong, the apex cuspidate-acuminate, the base rounded to acute; upper leaf surface glabrous, with major nerves both raised and channelled; lower surface glabrous (except for domatia), with all nerves raised; domatia (dense tufts of hispid hairs) present on lower leaf surface in the angles between lateral nerves and midrib, and also sometimes where lateral nerves meet. Venation: lateral nerves few (2–5 on each side), at an acute angle to the midrib; tertiary veins scalariform, ± at right angles to the midrib. Panicles axillary, many-flowered, loose; flowers greenish. Calyx 2–3 mm. long, lobes shortly connate at base, imbricate, elliptic, puberulous. Corolla tube c. 3 mm. long, barrel-shaped, pubescent externally, densely pilose internally except for a narrow glabrous band at base and apex, the hairs pointing downward; 5 small triangular scales present at mouth, one in each sinus between the petal lobes. Corolla lobes 3–5 mm. long, narrowly linear, ± glabrous. Stamens inserted almost at base of corolla tube; filaments short; anthers large (2–2·75 mm. long), reaching almost to mouth of corolla tube, each comprising a sagittate sterile structure bearing the anther thecae on its inner side on the upper half, the five anthers lightly connate by their margins into a tube round the gynoecium. Five square scales present on the receptacle centripetal to and alternating with stamens, reaching half-way up the ovary. Ovary of 2 ± free carpels c. 0·5 mm. long; style ± absent; clavuncle fusiform, style of 2 short fleshy lobes. Fruit a divaricate pair of follicles, each up to 25 cm. long and 1·3 cm. in diameter, narrowly cylindrical, tapering at both ends, brown-pubescent outside, smooth, glabrous and yellow within, splitting longitudinally at maturity, containing ∞ seeds. Seed c. 14 × 2·5 mm., narrowly ellipsoidal, strongly compressed, with apical coma of fine white simple hairs up to 3 cm. long.

Zambia. N: Luwingu Distr., Nsumbu I., L. Bangweulu, fl. 13.x.1947, *Brenan & Greenway* 8103 (K). **Zimbabwe.** E: Chipinge (Chipinga) Distr., Chirinda forest, 1100 m., fl. & fr. x.1964, *Goldsmith* 41/64 (BR; K; LISC; LISU; PRE; SRGH). S: Bikita Distr., Mt. Horzi, 1275 m., st.

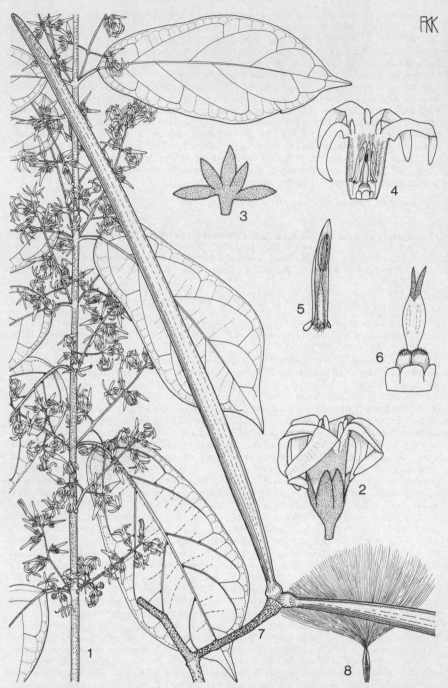

Tab. 118. ONCINOTIS TENUILOBA. 1, habit (×⅔); 2, flower (×4); 3, calyx, opened out (×4); 4, part of corolla, opened out (×4); 5, stamen (×12); 6, gynoecium with disc scales (×12), 1–6 from *Simão* 485; 7, fruit (×⅔), from *Drummond* 10187; 8, seed (×⅔), from *Wild* 2175.

11.v.1969, *Biegel* 3115 (K; SRGH; WAG). **Malawi.** S: Lisau, Chiradzulu F.R., c. 1400 m., fr. 14.i.1982, *Chapman & Balaka* 6095 (BR). **Mozambique.** N: Malema, 40 km. from Entre-Rios, road to Ribáuè, Murripa Mt., c. 1100 m., fr. 15.xii.1967, *Torre & Correia* 16521 (LISC; LISU). Z: Milange, Serra Tumbine, c. 1000 m., fr. 18.i.1966, *Correia* 450 (LISC). MS: Manica, Dombe, fl. 23.viii.1945, *Simão* 485 (LISC; LISU).

Also occurring in S. Africa (Cape Prov. and Natal), Tanzania, Kenya, Uganda, Ethiopia, Zaire, Congo (Brazzaville), Central African Rep., Nigeria and Cameroon. In forest.

29. BAISSEA A. DC.

Baissea A. DC., Prodr. **8**: 424 (1844).
Zygodia Benth. in Benth. & Hook. f., Gen. Pl. **2**: 716 (1876).
Perinerion Baill. in Bull. Soc. Linn. Paris **1**: 758 (1888).
Guerkia K. Schum. in Engl. & Prantl, Nat. Pflanzenfam. **4**, 2: 180, fig. 59 (1895).
Codonura K. Schum. in Engl., Bot. Jahrb. **23**: 229 (1896).

Rhizomatous creepers, climbing shrubs or lianas with white latex. Spines, tendrils and stipules absent. Leaves petiolate; petiole with glands on the adaxial side; lamina with the midrib impressed above and prominent beneath; secondary veins curved towards the margin, anastomosing; tufts of hairs (domatia) usually present in the axils of some secondary veins. Inflorescences axillary and sometimes at the same time terminal. Calyx lobes imbricate, almost free, at the base often with colleters inside. Corolla tube obconical, with a short urceolate base; inside with tufts of hairs alternating with the stamens; corolla lobes contorted in bud and overlapping to the right, spreading to reflexed. Stamens inserted almost at base of the corolla tube; anthers conivent in a cone around the gynoecium, adaxially pubescent, fertile in the upper part only, sagittate at the base, acuminate at the apex, coherent with the clavuncula by a retinacle (a circular patch between the tails inside on the connective 0·1–0·2 mm. in diam.). Ovary semi-inferior; carpels two, multi-ovulate, connate at the extreme base only, abruptly narrowed into the style, surrounded by a disk; disk adnate to the ovary at the base only, ring-shaped, with 5 oblong lobes alternating with the stamens; style obconical to almost cylindrical; clavuncula consisting of a slightly 5-winged upper and a cylindrical lower portion, to the latter of which the 5 retinacles adhere; stigma bilobed. Fruit composed of 2 pendulous, follicular, narrowly cylindrical mericarps, these sometimes united at the apices. Seeds narrowly ellipsoid, more or less laterally compressed, not rostrate, with an apical coma; endosperm in a thin layer completely surrounding the embryo.

A genus of 14 species, restricted to Africa.

1. Leaves subacute at the apex, often slightly mucronate; inconspicuous secondary veins at 5–25 each side; corolla lobes 7–22 mm. long, very narrowly triangular - 1. *wulfhorstii*
- Leaves acuminate at the apex, less often acute, rounded or emarginate; conspicuous secondary veins at 5–10 each side; corolla lobes 1–4·8 mm. long, triangular - - 2
2. Branchlets glabrescent; petiole 3–10 mm. long, glabrous or less often minutely pubescent at the base; leaves glabrous on both sides - - - - - - - 2. *viridiflora*
- Branchlets not glabrescent; petiole 1–3 mm. long, densely brown-pubescent; leaves pubescent on both sides, glabrescent except along secondary veins and margin
3. *myrtifolia*

1. **Baissea wulfhorstii** Schinz in Bull. Herb. Boiss. **4**: 816 (1896). — Stapf in F.T.A. **4**, 1: 207 (1902) tom. cit., Add.: 611 (1904). — White, F.F.N.R.: 347 (1962). — Codd in Fl. Southern Afr. **26**: 244, fig. 40 (1963). — H. Huber in Merxmüller, Prodr. Fl. S.W. Afr. **19**, 112, 3 (1967). TAB. **119**. Type from Namibia.
Baissea spectabilis Hua in Bull. Mens. Soc. Linn. Paris, N.S. **1**: 10 (1898). Type from Angola.

Rhizomatous creeper, climbing shrub, or liana up to 6 m. or more high, twining and climbing over shrubs and in trees. Branches purplish-grey, sometimes with small orange-brown lenticels; branchlets densely brown-pubescent. Leaves petiolate; petiole 0·5–3(6) mm. long, densely brown-pubescent; lamina elliptic to ovate, or narrowly so, (0·3)0·6–4·7 × (0·2)0·4–1·5 cm., at the apex subacute and often slightly mucronate, at the base cuneate, rounded or subcordate, chartaceous to subcoriaceous, pubescent on both sides, glabrescent except along the margin and on the midrib; at each side with 5–25 inconspicuous secondary veins. Flowers in 1–3-flowered, axillary lax cymes (often also in terminal 1–15-flowered, lax thyrsoid

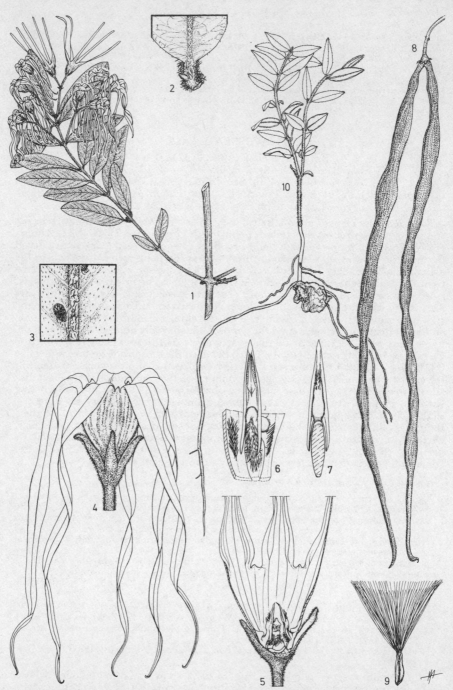

Tab. 119. BAISSEA WULFHORSTII. 1, flowering branch (×⅔), from *Murta & Silva* 723; 2, leaf base with petiole from above (×4), 3, domatia (×6), 2–3 from *Strid* 2381; 4, flower (×4), 5, flower opened up showing 2 stamens and the pistil the anther of the right stamen detached from the clavuncula, part of corolla, calyx and 3 stamens removed (×4), 6, stamen, abaxial side (×12), 7, stamen, adaxial side (×12), 4–7 from *Mutta & Silva* 723; 8, fruit (×⅔), 9, seed (×⅔), 8–9 from *Carr* 108; 10, seedling, 5 months (×1), from *Fanshawe* 7438.

inflorescences), sweet scented. Peduncle 5–30 mm. long, densely brown-pubescent. Sepals 1·8–5 mm. long, ovate to triangular, outside pubescent, inside pubescent except the base, ciliate. Corolla tube white outside, inside white with red longitudinal striations, 2·5–5 mm. long, pubescent outside except the portion covered by the sepals; inside with tufts of stiff hairs at 0·2–1·5 mm. from the base; lobes white, sometimes tipped with yellow, 7–22 mm. long, very narrowly triangular, sparsely pubescent or glabrous outside, often ciliate, inside glabrous. Stamens with filaments 0·2–0·5 mm. long, anthers 1·8–2·5 mm. long. Pistil 2·1–2·7 mm. long; carpels 0·8–1·2 mm. long; disk 0·1–0·2 mm. high and lobes c. 0·1 mm. long, pubescent; style 0·3–0·7 mm. long, pubescent; clavuncula 0·4–0·6 mm. long; stigma 0·2–0·4 mm. long. Fruit a pair of follicles, 15–50 cm. long and 2–8 mm. in diam., slightly constricted between the seeds, densely brown-pubescent outside. Seed 1·3–1·8 × 0·2–0·3 cm. with a coma 25–40 mm. long.

Botswana. N: 29·1 km. W. of Nokaneng, on road to Quangwa, fl. 23.x.1974, *Smith* 1154 (BR; K; MO; PRE; SRGH). **Zambia.** B: c. 16 km. N. of Senanga, c. 1000 m., fl. 3.viii.1952, *Codd* 7361 (BR; BM; COI; K; L; MO; NY; P; PRE; SRGH). N: Samfya, fl. 3.x.1953, *Fanshawe* 347 (BR; K). S: Livingstone Distr., Katambora, 2900 m., fl. 5.x.1955, *Gilges* 446 (K; LISU; PRE; SRGH).

Also in Angola and Namibia. In open woodland on sandy soil.

2. **Baissea viridiflora** (K. Schum.) De Kruif in Meded. Landb. Wag. **83**–7: 16 (1984). Type from Tanzania.

   Montandra viridiflora K. Schum. in Engl., Bot. Jahrb. **33**: 319 (1903). — Stapf in F.T.A. **4**, 1, Add.: 613 (1904). Type as above.

Liana up to 15 m. or more high, twining and climbing over shrubs and in trees. Branchlets pubescent, glabrescent. Leaves petiolate; petiole 3–10 mm. long, glabrous or less often minutely pubescent at the base; lamina elliptic or narrowly so, 2·5–11 × 1–4·6 cm., at the apex acuminate, less often acute, rounded or emarginate, cuneate at the base, glabrous on both surfaces; on each side 6–10 conspicuous secondary veins. Flowers aromatic, in 3–20-flowered axillary cymes, fairly congested, and (less often) also at the same time in a terminal thyrsoid, many flowered lax inflorescence. Peduncle 6–40 mm. long, pubescent. Sepals 0·6–1·2 mm. long, ovate to triangular, outside pubescent, inside pubescent except at the base, ciliate. Corolla tube white to greenish-yellow, 1·3–3 mm. long, outside glabrous to sparsely pubescent where exposed; inside with tufts of stiff hairs at 0·2–1·2 mm. from the base; lobes white to greenish-yellow, 1·8–4·8 mm. long, triangular, glabrous on both sides. Stamens with the filaments 0·3–0·4 mm. long; anthers 1·4–1·8 mm. long. Pistil 1·7–1·9 mm. long, carpels 0·9–1 mm. long; disk with a ring 0·1–0·2 mm. high and lobes c. 0·1 mm. long; style c. 0·2 mm. long; clavuncula c. 0·3 mm. long; stigma 0·2–0·3 mm. long. Follicles, 6–30 cm. long and 2–5 mm. in diam., often slightly constricted between the seeds, brown-pubescent outside, glabrescent. Seed 1·2–1·3 × 0·1 cm. with a coma 14–16 mm. long.

Malawi. S: Mt. Mulanje, Great Ruo Gorge, Lufiri, 770 m., fr. 18.vi.1962 *Richards* 16770 (K).

Also in Tanzania and Zaire. In rainforest.

3. **Baissea myrtifolia** (Benth.) Pichon in Bull. Mus. Nation. Hist. Nat. Paris, Sér. 2, **20**: 196 (1948). Type from Tanzania (Zanzibar).

   Zygodia myrtifolia Benth. in Hook: Ic. Pl. **12**: 73 (1876). — K. Schum. Engl., in Engl. & Prantl, Nat. Pflanzenfam. **4**, 2: 164 (1895); in Pflanzenw. Ost.-Afr. (Engl.) **C**: 318 (1895). — Stapf in F.T.A. **4**, 1: 218 (1902). Type as above.

   Zygodia urceolata Stapf in Kew Bull. **1894**: 122 (1894); in loc. cit. Type from Angola.

   Oncinotis melanocephala K. Schum. in Phys. Abh. Kön. Akad. Wiss. Berlin **1**: 34 (1894); in Engl. & Prantl, Nat. Pflanzenfam. **4**, 2: 179 (1895); tom. cit.: 319 (1895). Type from Tanzania.

   Zygodia kidengensis K. Schum. in Engl. & Prantl, Nat. Pflanzenfam. **4**, 2: 164 (1895) as "*kidengensis*"; in Pflanzenw. Ost.-Afr. (Engl.) **C**: 318 (1895) as "*kidengensis*". — Stapf in F.T.A. **4**, 1: 219 (1902). Type from Tanzania.

   Zygodia melanocephala (K. Schum.) Stapf in tom. cit.: 219 (1902). Type for *Oncinotis melanocephala*.

   Baissea urceolata (Stapf) Pichon, tom. cit.: 196 (1948) as "*kidengensis*". Type as for *Zygodia urceolata*.

   Baissea kidengensis (K. Schum.) Pichon, tom. cit.: 196 (1948) as "*kidengensis*". Type as for *Zygodia kidengensis*.

Baissea melanocephala (K. Schum.) Pichon in Bull. Jard. Bot. Brux. **24**: 35 (1954). Type as for *Oncinotis melanocephala*.

Climbing shrub or liana, up to 5 m. high or more, twining and climbing over shrubs and in trees, and up to 20 m. long or more. Trunk up to 2 cm. in diam. Branches grey to brown; branchlets densely brown-pubescent. Leaves petiolate; petiole 1–3 mm. long, densely brown-pubescent; lamina ovate to obovate, or narrowly so, (1)1·5–9·8 × (0·5)1–3·6 cm., at the margin revolute, at the apex acuminate, less often acute or emarginate, at the base rounded or cuneate, sometimes subcordate, pubescent on both surfaces, glabrescent except along secondary veins and margin; midrib pubescent on both surfaces, rarely glabrescent; on each side 5–10 conspicuous secondary veins. Flowers in 3–25-flowered axillary, fairly congested cymes; peduncle 0·5–16 mm. long, densely brown-pubescent. Sepals 1·1–3·8 mm. long, ovate, outside pubescent, inside pubescent except at the base, ciliate. Corolla tube outside yellow or sometimes orange, inside yellow to orange-red, 1·5–4 mm. long, glabrous outside, or pubescent except where exposed; inside with tufts of stiff hairs at 0·3–2·2 mm. from the base; lobes yellow, sometimes orange, 1–2·5 mm. long, triangular, glabrous outside often ciliate, inside at least at the base pubescent. Stamens with the filaments 0·3–0·7 m. long; anthers 1·5–2 mm. long. Pistil 0·2–0·8 mm. long, glabrous; carpels 0·7–1 mm. long; disk with a ring 0·1–0·2 mm. high and lobes c. 0·1 mm. long, pubescent; style 0·1–0·2 mm. long; clavuncula 0·2–0·5 mm. long; stigma 0·2–0·5 mm. long. Follicles, 7–44 cm. long and 2–5 mm. in diam., slightly constricted between the seeds, brown-pubescent outside. Seed 0·9–1·4 × 0·15–0·25 cm. with a coma 18–35 mm. long.

Zambia. N: Puta, fl. buds & fr. 18.viii.1958, *Fanshawe* 4718 (K).

Also occurring in Angola, Zaire, Congo (Brazzaville), Tanzania and Kenya. Riverine forest and thickets in woodland, often on rocky places.

INDEX TO BOTANICAL NAMES